RADAR PRECISION
AND RESOLUTION

G. J. A. Bird, C.Eng., M.I.E.R.E.
EMI Electronics Limited, Wells, Somerset

PENTECH PRESS LONDON

First Published 1974
Pentech Press Limited: London
8 John Street, WC1N 2HY

ISBN 0 7273 1801 2

Printed in England by
The Whitefriars Press Ltd., London and Tonbridge

UHI
Millennium
Institute

Please return/renew this item by the last date shown

Tillibh/ath-chlaraidh seo ron cheann-latha mu dheireadh

RADAR PRECISION
AND RESOLUTION

To Pat, David and Sarah

PREFACE

This book was written to help practising engineers, and students of radar theory, to understand how uncertainty function techniques can be used to analyse the performance of radar systems. It is common for modern systems to employ sophisticated modulation, and their characteristics are not always apparent from an intuitive treatment.

Engineers must have confidence in mathematical techniques (and knowledge of any limitations) if they are to apply them to new problems. It is the author's opinion that this confidence is best inspired by stressing basic principles and by giving the proofs of all mathematical results, thereby establishing the conditions necessary for their validity. This philosophy has been adopted in writing the book. To facilitate a first reading, the conclusions and discussion usually precede the detailed mathematics; this also makes it easy to use the book for reference purposes.

The book has been divided into two parts. Chapters 1 to 5 cover the radar theory and include numerous references to Chapters 6 and 7 which provide a firm foundation of the underlying transform theory. It is suggested that the reader may like to concentrate initially on Chapter 1, followed by Sections 4.1, 4.2, 4.6 and Chapter 5. This procedure will acquaint him with the significance of the uncertainty function and how it can be used to divide radar receivers into two classes—matched filter receivers and Fourier transform receivers. The theory is consolidated in Chapter 5 by applying the results to two practical modulation waveforms; the reader should then be in a position to understand the more advanced results given in the established literature.

The second part of the book provides the mathematical background to Chapters 1 to 5. Chapter 6 deals with the relationship between Laplace and Fourier transforms. The discrete Fourier transform (DFT) is also discussed and it is shown how two DFT theorems lead to the fast Fourier transform (FFT) process. Chapter 7 covers the Hilbert transform and complex analytic signals. The author has always found 'engineering' treatments of this latter subject to be particularly unconvincing—it is hoped that the reader will find the present treatment to be much more satisfactory. The author has found Part 2 to be a useful reference source when working in other fields of signal processing. The summary of the main notation in Appendix 5 is also useful in this respect.

Preface

The first draft of this book was written as an internal report for the Wells laboratories of EMI Electronics Ltd. The material has also been used as the basis of a series of after-hours lectures to the author's colleagues. The author would like to thank EMI Electronics Ltd. for permission to publish and also his colleagues for many helpful discussions.

G. J. A. Bird

CONTENTS

Contents

Chapter 1.

THE RADAR UNCERTAINTY FUNCTION

The uncertainty function, introduced by Woodward [1], is a function of two variables representing signal delay and Doppler shift. It may be regarded as a 'figure of merit' by which various waveforms and processing methods can be compared. This chapter gives a qualitative treatment of the uncertainty function and its properties.

1.1 LIMITATIONS AND CHARACTERISTICS OF THE UNCERTAINTY FUNCTION

The return signal from a radar target is a modified version of the transmitted signal. The modifications are due to the parameters of the target which can, in principle, be deduced by comparing the returned signal with the transmitted signal.

The uncertainty function results from a consideration of the signal returned from a single point target. An example of such a target would be a small insect flying at constant velocity toward an 'upward looking' radar.

The radar return is assumed to differ from the transmitted signal in only two ways, namely:

(1) A time delay (x), proportional to the radial range of the target.
(2) A constant frequency shift (y) of the whole signal spectrum, proportional to the target radial velocity. The variable y is called the Doppler offset; it is positive for targets travelling towards the radar.

Strictly speaking, the assumption of a constant Doppler offset is an approximation which is only valid when the target velocity is small compared with the velocity of propagation of the transmitted signal. The true effect is a distortion of the signal spectrum, due to an offset which varies with frequency.

It is normally realistic to assume a constant Doppler offset in the case of electro-magnetic propagation in air, but caution must be exercised in such applications as sonar [2].

With the above assumptions, the target may be regarded, at any one time, as a point in the x, y plane. Use of the uncertainty function allows one to define an area of uncertainty in the x, y plane, inside

1

which the target may lie 'Precision' describes the smallness of this area. Some signals lead to multiple areas of uncertainty for a single target; such signals are said to exhibit ambiguities.

The uncertainty function is defined in a manner which takes no account of such factors as range attenuation and aerial polar diagram, since they are not functions of the transmitted waveform. If necessary, these considerations can be incorporated as weighting along the x-axis, and will often eliminate areas of ambiguity. In a similar fashion, other ambiguous areas can often be eliminated by a knowledge of the maximum target velocity.

The above remarks can be summarised and amplified by a list of some of the characteristics of the uncertainty function as follows:

(1) It is often possible to specify a radar target completely by means of two characteristics of the returned signal, namely, delay time (x) and Doppler offset (y). If the above is true, the target could be represented as a spike in the x, y plane, the height of the spike being proportional to the target amplitude. The uncertainty function is calculated from the parameters of the transmitted signal, and can be visualised as a 3-dimensional solid centred on the x, y co-ordinates of the target.

Replacing the 'spike' of the target by an appropriately scaled uncertainty function shows up the inherent precision of the transmitted waveform; i.e., the radar does not 'see' the target as a single point.

(2) The formula giving the uncertainty function of a particular waveform will show how the various waveform parameters may be modified to minimise any undesirable characteristics regarding precision. For simple waveforms the necessary modifications are often obvious (e.g. a short a.m. pulse gives better range precision than a long one); for a complicated waveform they may not be at all obvious. The uncertainty function provides a unified method of signal analysis.

(3) An arbitrary signal-processing method does not necessarily exploit all the precision inherent in a signal, e.g. the high range-precision inherent in a short a.m. pulse is lost if the receiver bandwidth is not wide enough. It is useful to judge signal-processing schemes by comparing their output waveforms with appropriate cuts through the uncertainty function.

(4) Two signal-processing schemes which retain all the precision inherent in a signal are discussed in Chapter 4. One scheme, the 'matched filter receiver', has an output waveform which is a cut through the uncertainty function, parallel to the delay axis; the other scheme, the 'Fourier transform receiver', gives a cut parallel to the Doppler axis.

(5) The word 'precision' is used to describe the radar performance when used against a single target, while 'resolution' refers to the performance against multiple targets and clutter. The uncertainty function only gives information regarding the inherent precision of a waveform; it does not describe the inherent resolution.

(6) Property (5) should be interpreted positively. If a receiver is proposed having the features of (4) above, no further signal processing will improve its precision qualities. It may happen that the uncertainty function has sidelobes which decrease the resolution of a small target which is close to larger targets. There is nothing 'in the rules' to say that further signal processing will not reduce the undesirable sidelobes; it will frequently be possible to do this, the price usually being a broadening of the main lobe. Thus, although modifications of the 'optimum' receivers of (4) are usually necessary, the uncertainty function still provides a reference against which any lessening of precision may be judged.

1.2 THE RMS ERROR CRITERION

Assume that the transmitted signal is of the form

$$f(t, \omega_0) = |a(t)| \cos(\omega_0 t + \phi(t))$$

A return signal with delay (x) and Doppler offset (y) will be given by

$$f(t-x, \omega_0 + y)$$

As discussed in Section 1.1, such a signal can be represented as a point in the x, y plane. One can only be sure that the point is situated at x_1, y_1 as opposed to $(x_1 - \tau), (y_1 + \omega)$, if

$$f(t-x_1, \omega_0 + y_1)$$

is sufficiently different from

$$f(t-x_1 + \tau, \omega_0 + y_1 + \omega)$$

The method of defining the difference between two signals cannot be uniquely specified. Any measure is suitable as long as it gives meaningful results. Whether the signals are different enough to be distinguished depends upon an agreed threshold level, which the measure of the difference must have for distinguishability. Different measures will lead to different thresholds, but the same conclusions.

The measure of difference which leads to the uncertainty function is the r.m.s. difference, $\&$. Thise measure is used frequently in mathematical and control theory literature, mainly because of its convenient mathematical properties. For the present discussion, $\mathscr{E}$ will be defined through the following relationship:

$$\mathscr{E}^2 = \int_{-\infty}^{\infty} \{f(t-x, \omega_0 + y) - f(t-x+\tau, \omega_0 + y + \omega)\}^2 \, dt \qquad (1.1)$$

It is shown in Section 2.1 that the above definition leads to

$$\mathscr{E}^2 = 2 \int_{-\infty}^{\infty} [f(t)]^2 \, dt - |\chi(\tau, \omega)| \cos[(\omega_0 + y + \omega)\tau - \text{Arg}\{\chi(\tau, \omega)\}] \qquad (1.2)$$

where

$$\chi(\tau, \omega) = \int_{-\infty}^{\infty} a(t)a^*(t+\tau) \, e^{-j\omega t} \, dt \qquad (1.3)$$

and

(1) $a(t)$ is a complex baseband function which is derived from the transmitter modulation, as discussed in Section 2.1.
(2) τ, ω are distances in the x, y plane measured with respect to the target co-ordinates (see Fig. 1.2).
(3) $\int_{-\infty}^{\infty} [f(t)]^2 \, dt = E$ (i.e. the energy of the transmitted signal).

In this book $|\chi(\tau, \omega)|$ is called 'the uncertainty function' although in the established literature it is sometimes referred to as 'the ambiguity function'. Also, $\chi(\tau, \omega)$ and $|\chi(\tau, \omega)|^2$ can be found called by either name. A full mathematical treatment of $\chi(\tau, \omega)$ is given in Chapter 2.

Only two properties are relevant at this stage.

(1) $\chi(\tau, \omega)$ is never greater than $\chi(0, 0)$, its value at the origin; at which point it is equal to $2E$. Hence the ratio $\chi(\tau, \omega)/2E$ is independent of signal energy, E.
(2) $\chi(\tau, \omega)$ normally changes with τ much more slowly than does $\cos[(\omega_0 + y + \omega)\tau]$ because ω_0 is a high carrier frequency.

Equation 1.2 can be re-written in the more convenient form:

$$\frac{\mathscr{E}^2}{2E} = 1 - \frac{1}{2E} |\chi(\tau, \omega)| \cos[(\omega_0 + y + \omega)\tau - \text{Arg}\{\chi(\tau, \omega)\}] \qquad (1.4)$$

A graphical interpretation of Equation 1.4 is given in Fig. 1.1 which shows a typical variation of $\mathscr{E}^2/2E$ along a line in the τ, ω plane parallel to the τ axis. E_t represents an arbitrary threshold which $\mathscr{E}^2$ must exceed before the signal can be said to be distinguishable from one with different τ, ω co-ordinates.

As ω_0 is a high carrier frequency, $\mathscr{E}^2/2E$ is a rapidly fluctuating function of τ, and always lies between the solid lines in Fig. 1.1.

Since precision with ambiguities separated by distances of the order of the transmitted signal wavelength is not normally useful, the lower

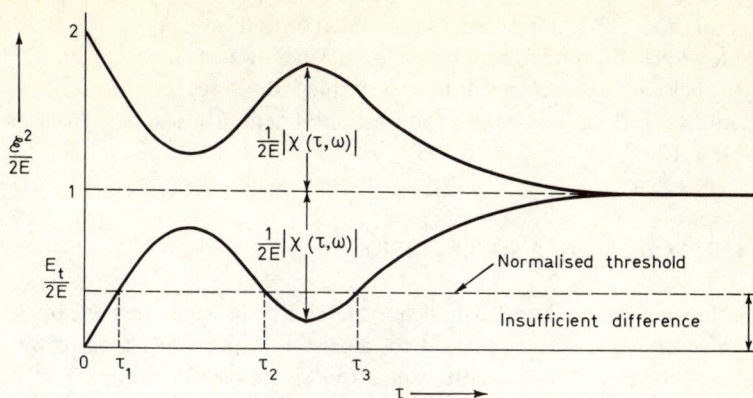

Fig. 1.1. A graphical interpretation of Equation 1.4

(minimum value) curve is chosen as a test of whether the threshold has been exceeded. Thus, the target range co-ordinate can be said to lie in the ranges 0 to τ_1, or τ_2 to τ_3. The example used has both range-uncertainty and ambiguity.

Figure 1.1 shows that achievable precision can be increased by reducing the ratio E_t/E. Since E_t is ultimately set by system noise, this implies an increase in the transmitted energy, E.

1.3 THE AREA OF UNCERTAINTY OF A RECEIVED SIGNAL

Figure 1.1 shows that Equation 1.4 can be re-written to read:

A received signal may be located where $\dfrac{|\chi(\tau, \omega)|}{2E} \geqslant 1 - E_t/2E$ (1.5)

E_t in Equation 1.5 represents the (arbitrary) threshold level which the mean square error must exceed before it is possible to measure the parameters of the returned signal. Since system noise has not been considered, the implication is that E_t exceeds the system noise level by a significant amount.

Table 1.1

E/E_t (ratio)	Contour of $|\chi(\tau, \omega)|/2E$ which defines the area of uncertainty
1	−6 dB
2	−2.5 dB
10	−0.4 dB
100	−0.04 dB

$|\chi(\tau, \omega)|/2E$ has been converted to dB by the formula: $20 \log_{10} r$.

Equation 1.5 can be used to calculate contour levels of $|\chi(\tau, \omega)|/2E$ inside which the target parameters of a signal of a given energy lie. The area enclosed by such a contour is termed the 'area of uncertainty'. Examples for various signal-to-threshold energy ratios are given in Table 1.1.

1.4 RESOLUTION AND PRECISION

Resolution is used here to indicate the ability to separate the returns from several targets (as opposed to 'precision' which was used above to indicate the performance with respect to a single target).

If the required precision has been defined as an area in the τ, ω plane, and that area is enclosed within a much larger area of uncertainty (for a given signal-to-threshold ratio) it can be stated that the given signal does not meet the specification. No amount of clever signal processing will alter this fact (assuming, of course, that the signal-energy to threshold-level cannot be increased).

The uncertainty function cannot be used directly to make a similar statement regarding resolution.

If the uncertainty areas corresponding to two targets are plotted in the x, y plane and they overlap, resolution will be difficult or impossible. If the two signals were much higher than the receiver threshold level the corresponding uncertainty areas would be small and might not overlap. However, one signal could still be much stronger than the other and its response sidelobes could completely swamp the weaker signal. One could only be sure that this would not happen if the $|\chi| = 0$ contours surrounding each target co-ordinate did not intersect. This latter criterion is not very useful for the types of signal which are usually studied by means of the uncertainty function, and does not help in the case of a signal in clutter.

Intuitively, one might think that if the area of uncertainty of the weakest signal alone was marked off in the x, y plane the test for resolution in the presence of a signal n dB stronger would be whether the $-n$ dB contour of $|\chi|/2E$, surrounding the stronger signal co-ordinates, enclosed the area of uncertainty of the weaker signal. Although this argument has validity in the case of receivers which give an output waveform in the shape of a cut through $\chi(\tau, \omega)$, it does not apply to cases where this is not so. In particular, such alternative cases may well lead to better resolution.

A true assessment of the precision and resolution properties of a given radar system can only be made by studying the shape of the processed output pulse which is used to determine the value of the delay or Doppler parameter being measured. If a single figure is required as a measure of precision or resolution, it can be obtained by measuring the width of this output pulse at some arbitrary level.

In this book the 'precision' of a system will be defined as the width of its output pulse measured between its half-amplitude (i.e. -6 dB) points. Table 1.1 shows that this implies a signal-to-threshold energy ratio of unity. The 'n dB resolution' of a system will be defined as the width of its output pulse measured between its widest $-n$ dB points.

The above definitions apply to an actual system, the *inherent* precision of a transmitted signal will be defined as the -6dB width of the appropriate cut through $|\chi(\tau, \omega)|$.

τ and ω are related to range and velocity by the formulae

$$\tau = \frac{2r}{c} \tag{1.6}$$

$$\omega = \frac{2\omega_0 v}{c} \tag{1.7}$$

where r and v are range and velocity (with respect to the target), c is the velocity of propagation and ω_0 is the carrier frequency.

Thus if, for a given system, the delay precision is given by τ_1, the corresponding range precision will be $(c\tau_1/2)$. The velocity precision, corresponding to a Doppler precision of ω_1, will be $(c\omega_1/2\omega_0)$ and it should be noted that this can be improved by increasing ω_0.

1.5 THE RELATIONSHIP BETWEEN THE τ, ω AND x, y PLANES
The co-ordinates x, y define a point with respect to the receiver, while the co-ordinates τ, ω define the same point with respect to the target, Fig. 1.2.

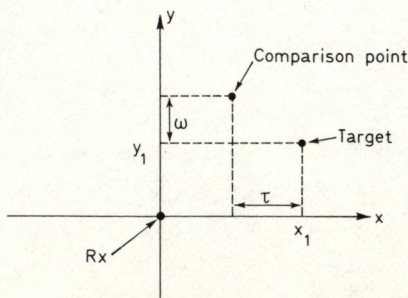

Fig. 1.2. Relationship between x, y and τ, ω

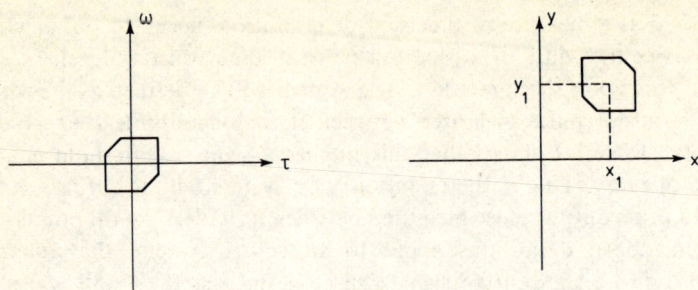

Fig. 1.3. Translation of $\chi(\tau, \omega)$ to x, y plane

Since

$$x = (x_1 - \tau), \qquad y = (y_1 + \omega)$$

it follows that

$$\chi(\tau, \omega) = \chi[-(x - x_1), (y - y_1)]$$

The above relationship is illustrated by Fig. 1.3.

Chapter 2

THE MATHEMATICAL TREATMENT OF THE UNCERTAINTY FUNCTION

This chapter is used to show how the r.m.s. error criterion leads to the function

$$\chi(\tau, \omega) = \int_{-\infty}^{\infty} a(t)a^*(t+\tau) e^{-j\omega t} dt.$$

Also, some mathematical properties of $\chi(\tau, \omega)$ are studied.

2.1 DERIVATION OF THE UNCERTAINTY FUNCTION

It is assumed that a real signal, $f(t)$, is transmitted such that

$$f_a(t) = a(t) e^{j\omega_0 t} \qquad (2.1)$$

is the corresponding complex analytic signal (Chapter 7). Equation 2.1 implies that the real transmitted signal is given by

$$f(t) = \text{Re}\{f_a(t)\} \qquad (2.2)$$

Defining

$$a(t) = |a(t)| e^{j\phi(t)} \qquad (2.3)$$

leads to

$$f(t) = |a(t)| \cos[\omega_0 t + \phi(t)] \qquad (2.4)$$

A subtle point worth noting is that with the above definition of $f(t)$, $|a(t)|$ and $\phi(t)$ are not arbitrary functions. Let $f(t)$ be written:

$$f(t) = \text{Re}\{b(t) e^{j\omega_0 t}\}$$

where

$$b(t) = |b(t)| e^{j\theta(t)}$$

that is

$$f(t) = |b(t)| \cos[\omega_0 t + \theta(t)] \qquad (2.5)$$

where $|b(t)|$, $\theta(t)$ are defined as the real, arbitrary, functions applied to the amplitude and phase modulation terminals, respectively, of the transmitter.

9

Unless the carrier frequency ω_0 is sufficiently high, it does not follow that $a(t)$ is equal to $b(t)$. Rather, $a(t)$ has to be calculated from Equation 2.5 using the relationship:

$$a(t)\, e^{j\omega_0 t} = f_a(t) = f(t) + j\hat{f}(t) \qquad (2.6)$$

A full discussion of this point is given in Section 7.4. The distinction between the complex analytic signal and the exponential approximation has to be made in the case of sonar. Some important differences between radar and sonar are detailed by Kramer [2].

Fortunately, for practical radar applications, it is not necessary to use Equation 2.6 as $a(t)$ can usually be considered equal to $b(t)$.

The r.m.s. error criterion has been discussed in Section 1.2. The r.m.s. difference, $\mathcal{E}$, between two real signals $f(t)$ and $g(t)$, is defined through the relationship

$$\mathcal{E}^2 = \int_{-\infty}^{\infty} [f(t) - g(t)]^2 \, dt \qquad (2.7)$$

An important property of complex analytic signals (proved in Section 7.2.5) is that Equation 2.7 can be replaced by

$$\mathcal{E}^2 = \tfrac{1}{2} \int_{-\infty}^{\infty} |f_a(t) - g_a(t)|^2 \, dt \qquad (2.8)$$

Using the relationships

$$f_a(t) = f(t) + j\hat{f}(t)$$

$$g_a(t) = g(t) + j\hat{g}(t)$$

Equation 2.8 can be expanded to give

$$2\mathcal{E}^2 = \int_{-\infty}^{\infty} |f_a(t)|^2 \, dt + \int_{-\infty}^{\infty} |g_a(t)|^2 \, dt - 2 \int_{-\infty}^{\infty} \mathrm{Re}\{f_a(t) g_a^*(t)\} \, dt \qquad (2.9)$$

Equation 2.9 follows since

$$\mathrm{Re}\{f_a(t) g_a^*(t)\} = f(t)g(t) + \hat{f}(t)\hat{g}(t)$$

To prove Equation 1.2 it is necessary to put

$$f(t) = f[t - x, \, \omega_0 + y]$$

$$g(t) = f[t - x + \tau, \, \omega_0 + y + \omega]$$

Hence

$$f_a(t) = a(t - x)\, e^{j(\omega_0 + y)(t - x)}$$

$$g_a(t) = a(t - x + \tau)\, e^{j(\omega_0 + y + \omega)(t - x + \tau)}$$

Substitution of the above in Equation 2.9 gives

$$2\mathcal{E}^2 = \int_{-\infty}^{\infty} |a(t-x)|^2 \, dt + \int_{-\infty}^{\infty} |a(t-x+\tau)|^2 \, dt$$

$$-2 \int_{-\infty}^{\infty} \mathrm{Re}\{a(t-x)a^*(t-x+\tau) \, e^{-j\omega t} e^{j\omega x} e^{-j(\omega_0+y+\omega)\tau}\} \, dt \quad (2.10)$$

By using the relationship between the spectra of real and complex analytic signals (Section 7.1), and by the application of Parseval's theorem (Appendix 2), it can be shown that

$$\int_{-\infty}^{\infty} |a(t-x)|^2 \, dt = \int_{-\infty}^{\infty} |a(t-x+\tau)|^2 \, dt = \int_{-\infty}^{\infty} |f_a(t)|^2 \, dt = 2 \int_{-\infty}^{\infty} [f(t)]^2 \, dt$$

Also, by changing the dummy variable t to $(t + x)$, the last integral in Equation 2.10 becomes

$$2 \, \mathrm{Re}\left\{ e^{j\omega x} e^{-j(\omega_0+y+\omega)\tau} \int_{-\infty}^{\infty} a(t)a^*(t+\tau) \, e^{-j(t+x)\omega} \, dt \right\}$$

Thus, with the definition

$$\chi(\tau, \omega) = \int_{-\infty}^{\infty} a(t)a^*(t+\tau) \, e^{-j\omega t} \, dt \quad (2.11)$$

Equation 2.10 may be written in the form

$$\mathcal{E}^2 = 2 \int_{-\infty}^{\infty} [f(t)]^2 \, dt - \mathrm{Re}\{e^{-j(\omega_0+y+\omega)\tau} \chi(\tau, \omega)\} \quad (2.12)$$

Equation 2.12 is also equivalent to

$$\mathcal{E}^2 = 2 \int_{-\infty}^{\infty} [f(t)]^2 \, dt - |\chi(\tau, \omega)| \cos[(\omega_0+y+\omega)\tau - \mathrm{Arg}\,\chi(\tau, \omega)] \quad (2.13)$$

The discussion in Section 1.2 uses Equation 2.13 to establish the physical significance of $\chi(\tau, \omega)$.

2.2 THE MATHEMATICAL PROPERTIES OF $\chi(\tau, \omega)$

$\chi(\tau, \omega)$ is defined by

$$\chi(\tau, \omega) = \int_{-\infty}^{\infty} a(t)a^*(t+\tau) \, e^{-j\omega t} \, dt \quad (2.14)$$

The most important properties of $\chi(\tau, \omega)$ are stated below; proofs are given in Section 2.5. Other properties can be found by reference to Stutt [3, 4] and Siebert [5].

$$\chi(\tau, \omega) = \mathscr{F}\{a(t)a^*(t+\tau)\} \quad (2.15)$$

$$\chi(\tau, \omega) = \frac{1}{2\pi} \int_{-\infty}^{\infty} A^*(jx) A\left[j(\omega + x)\right] e^{-jx\tau} dx \qquad (2.16)$$

$$\chi^*(\tau, \omega) = e^{-j\omega\tau} \chi(-\tau, -\omega) \qquad (2.17)$$

$$|\chi(\tau, \omega)| = |\chi(-\tau, -\omega)| \qquad (2.18)$$

$|\chi(\tau, \omega)|$ can be regarded as a 3-dimensional solid, placed upon the τ, ω plane. Two-dimensional functions can be obtained by taking cuts through $|\chi(\tau, \omega)|$. The symmetry indicated in Equation 2.18 means that the resulting functions will be even, if the cuts pass through the origin of the τ, ω plane.

$$\chi(0, 0) = \int_{-\infty}^{\infty} |a(t)|^2 dt \geqslant |\chi(\tau, \omega)| \qquad (2.19)$$

Equation 2.19 means that $|\chi(\tau, \omega)|$ can never be greater than its value at the origin. Note that since $a(t)$ was derived from $f_a(t)$

$$\int_{-\infty}^{\infty} |a(t)|^2 dt = 2 \int_{-\infty}^{\infty} [f(t)]^2 dt;$$

i.e. $\chi(0, 0)$ is equal to twice the energy of $f(t)$.

$$\frac{1}{2\pi} \iint_{-\infty}^{\infty} |\chi(\tau, \omega)|^2 d\tau \, d\omega = [\chi(0, 0)]^2 \qquad (2.20)$$

Equation 2.20 expresses the fact that the total volume of the $|\chi(\tau, \omega)|^2$ solid is constant, regardless of the form of $a(t)$. This means that any steps taken to concentrate $|\chi(\tau, \omega)|^2$ in a narrow spike at the origin must also result in a large spread of lower levels of the function. A change in the form of $a(t)$ gives the results set out in Table 2.1

Table 2.1

$a(t)$	$A(j\omega)$	$\chi(\tau, \omega)$	
$a(t) e^{jbt^2}$		$e^{-jb\tau^2} \chi[\tau, \omega + 2b\tau]$	(2.21)
	$e^{jb\omega^2} A(j\omega)$	$e^{jb\omega^2} \chi[\tau - 2b\omega, \omega]$	(2.22)
$a(\alpha t)$		$\frac{1}{|\alpha|} \chi[\alpha\tau, \omega/\alpha]$	(2.23)
	$A(j\alpha\omega)$	$\frac{1}{|\alpha|} \chi[\tau/\alpha, \alpha\omega]$	(2.24)

The significance of Equation 2.21 is discussed in Section 2.4.

Repetition of a basic waveform, i.e. changing $a(t)$ to

$$\sum_{i=0}^{n-1} C_i a(t - ik)$$

leads to an uncertainty function given by

$$\sum_{m=1}^{n-1} e^{-j\omega mk} \chi(\tau+mk, \omega) \sum_{i=0}^{n-1-m} C_i^* C_{i+m} e^{-j\omega ik}$$

$$+ \sum_{m=0}^{n-1} \chi(\tau-mk, \omega) \sum_{i=0}^{n-1-m} C_i C_{i+m}^* e^{-j\omega ik} \qquad (2.25)$$

where C_i is a real or complex multiplier (commonly unity) and n the number of pulses. The significance of Equation 2.25 is discussed in Section 2.3.

The composite function $a(t) + b(t)$, leads to

$$\chi(\tau, \omega) = \chi_{aa}(\tau, \omega) + \chi_{bb}(\tau, \omega) + \chi_{ab}(\tau, \omega) + e^{j\omega\tau}\chi_{ab}^*(-\tau, -\omega) \qquad (2.26)$$

where

$$\chi_{uv}(\tau, \omega) = \int_{-\infty}^{\infty} u(t)v^*(t+\tau)\,e^{-j\omega t}\,dt \qquad (2.27)$$

or, by inspection

$$\chi_{uv}(\tau, \omega) = \mathscr{F}\{u(t)v^*(t+\tau)\}\,dt \qquad (2.28)$$

2.3 THE EFFECT OF REPETITION UPON THE UNCERTAINTY FUNCTION

A general case of repetition of a basic waveform, $a(t)$, n times is given by

$$b(t) = \sum_{i=0}^{n-1} C_i a(t-ik)$$

By letting C_i take on complex values, i.e.

$$C_i = a_i\,e^{j\phi_i}$$

succeeding pulses can be given different amplitudes and initial phases.

The resultant uncertainty function is given by Equation 2.25. The proof of Equation 2.25 is given in Section 2.5.9.

Three important cases arise where the amplitudes of succeeding pulses remain constant and either

(1) The modulation is applied by 'gating' a continuously-running carrier oscillator, i.e.

$$f(t) = \sum_{i=0}^{n-1} |a(t-ik)|\cos[\omega_0 t + \phi(t-ik)]$$

(2) The oscillator is restarted for each 'pulse', and succeeding r.f. pulses are identical, i.e.

$$f(t) = \sum_{i=0}^{n-1} |a(t-ik)| \cos[\omega_0(t-ik) + \phi(t-ik)]$$

(3) The oscillator is restarted for each 'pulse', but the initial phase is random, i.e.

$$f(t) = \sum_{i=0}^{n-1} |a(t-ik)| \cos[\omega_0 t + \phi(t-ik) + \phi_i]$$

It is shown in Section 2.5.10 that the complete χ function for case (1) becomes

$$\sum_{m=-(n-1)}^{(n-1)} e^{-j(n-1-m)\theta} \frac{\sin[(n-|m|)\theta]}{\sin\theta} \chi[\tau-mk, \omega]$$

where

$$\theta = \frac{\omega k}{2}$$

For case (2), the χ function is given by a similar expression

$$\sum_{m=-(n-1)}^{(n-1)} e^{j\omega_0 km} e^{-j(n-1-m)\theta} \frac{\sin[(n-|m|)\theta]}{\sin\theta} \chi[\tau-mk, \omega].$$

For case (3) the χ function is given by

$$\sum_{m=1}^{n-1} e^{-j\omega mk} \chi(\tau+mk, \omega) \sum_{i=0}^{n-1-m} e^{j(\phi_{i+m}-\phi_i)} e^{-j\omega ik}$$

$$+ \sum_{m=1}^{n-1} \chi(\tau-mk, \omega) \sum_{i=0}^{n-1-m} e^{j(\phi_i-\phi_{i+m})} e^{-j\omega ik}$$

$$+ e^{-j(n-1)\theta} \left\{ \frac{\sin[n\theta]}{\sin\theta} \right\} \chi(\tau, \omega).$$

Assuming, for simplicity, that $\chi(\tau, \omega)$ of a single pulse is contained in a finite area of the τ, ω plane (Fig. 2.1) the effect of repetition is to give a new $\chi(\tau, \omega)$ consisting of $(2n - 1)$ of the original functions spaced along the τ axis at intervals of k (the repetition period).

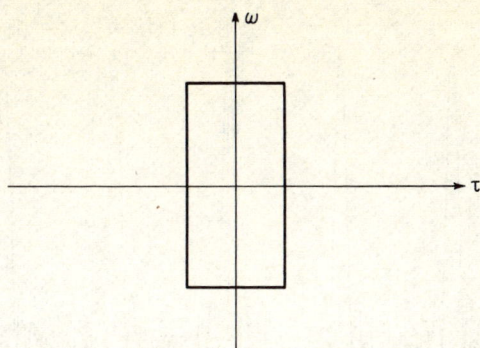

Fig. 2.1. Total area of uncertainty of a hypothetical $\chi(\tau, \omega)$

For cases (1) and (2) all the multiple functions are weighted by the factor

$$\frac{\sin\left[\dfrac{\omega k}{2}(n-|m|)\right]}{\sin\left[\dfrac{\omega k}{2}\right]}$$

where $|m|$ takes on values between 0 and $(n-1)$, according to which one of the multiple functions is being considered. For case (3), only the function corresponding to $m = 0$ is affected by the weighting factor.

Plots of the weighting factor are shown in Figs 2.2 and 2.3. The independent variable is taken as m (effectively τ) and ω, respectively.

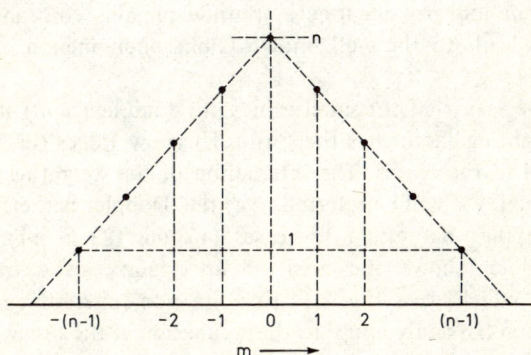

Fig. 2.2. $\dfrac{\sin\left[\dfrac{\omega k}{2}(n-|m|)\right]}{\sin\left[\dfrac{\omega k}{2}\right]}$ *for* $\omega = 0$ *and* m *an integer*

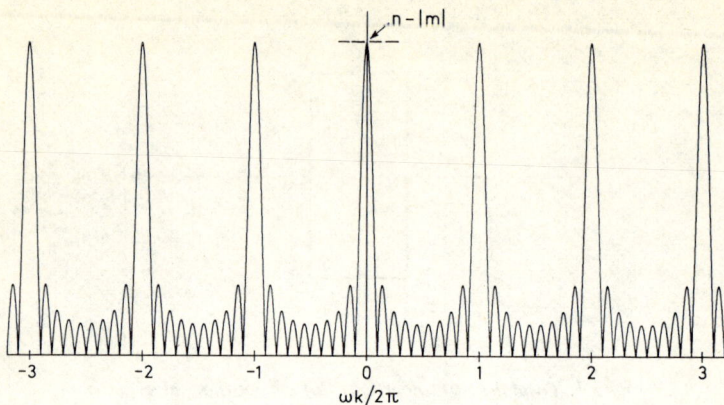

$$\text{Fig. 2.3.} \quad \left| \frac{\sin\left[\frac{\omega k}{2}(n-|m|)\right]}{\sin\left[\frac{\omega k}{2}\right]} \right| \quad \text{for } n-|m| \text{ high}$$

The function

$$f(x) = \frac{\sin(nx)}{\sin x}$$

is discussed fully by Guillemin [7] where it is shown that, for large n, it has the form

$$n\left[\frac{\sin(nx)}{nx}\right]$$

repeated when x equals integral multiples of π. As n gets larger, the main lobes of the above function become narrower and taller, but the ratio of main-lobe to side-lobe amplitude remains constant; it is this fact which leads to the well known Gibbs phenomenon of spectrum theory.

It can be seen that for small values of τ (and hence m) and for large n, the weighting factor has the form of narrow ridges (or bars) which run parallel to the τ axis. The interaction of the weighting factor with the individual 'χ's' will be referred to as the 'Doppler bar' effect.

The Doppler bar effect for cases (1) and (2) is summarised by Fig. 2.4 which shows the area of uncertainty of a train of the waveforms which gave Fig. 2.1. For the non-coherent case (3), the concentration bars only apply to the χ function at the origin.

Using the quantitative definition of precision given in Section 1.4 means that the Doppler precision resulting from the Doppler bar effect is given by

$$\left(\frac{\omega}{2\pi}\right)nk = 1.2, \qquad n \gg 1$$

The function which is of interest, from the point of view of radar theory, is $|\chi(\tau, \omega)|$ rather than $\chi(\tau, \omega)$. Figure 2.4 illustrates that, in regions of the τ, ω plane where no overlap occurs, the modulus of the combined function is equal to the modulus of the appropriate 'χ', times the weighting factor.

The multiple areas of uncertainty produced by repetition are mathematical expressions of the well known 'second (or higher)—time-round effects' encountered in practical radar systems. In practice, the multiple areas will also be weighted by the effects of range attenuation. One factor in the choice of pulse repetition frequency ($1/k$) is to remove ambiguities and allow the 'χ' function at the origin of Fig. 2.4 to be the dominant one. If the latter condition is satisfied, it can be stated that the Doppler bar effect applies to cases (1), (2) and (3), i.e. gated, pulsed and non-coherent sources.

The fact that the amplitude of the total uncertainty function increases as n increases means that, for a given receiver threshold, repetition increases the potential precision. Physically, this is because the energy of a repetitive train is n times the energy of a single pulse. When plotting the uncertainty function, for different values of n, it is convenient to remove the above effect by dividing by n.

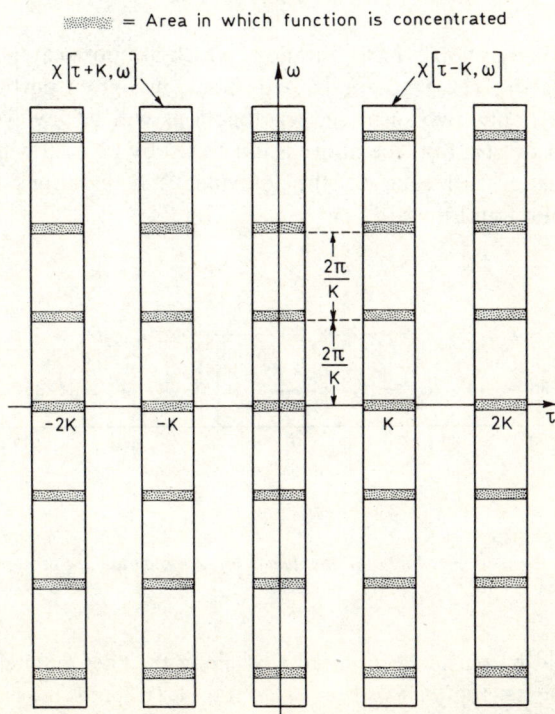

Fig. 2.4. Total area of uncertainty of a train of the pulses which gave Fig. 2.1

2.3.1 Coded waveforms

The full form of Equation 2.25 has to be used when studying deliberately coded wavetrains. In this case no Doppler bar effect will occur (the effect of coded repetition is to produce a new 'single-word' waveform which is made up of a number of bits). Of course if a train of identical words is transmitted, the Doppler bar effect will act upon the uncertainty function of a single word.

When considering the uncertainty function of a word consisting of L contiguous coded bits, occurring at $t = 0$, $t = d$, $t = 2d$, etc. it is convenient to write Equation 2.25 in the following form:

$$\chi(\tau, \omega) = \sum_{m=0}^{L-1} \chi_b(\tau - md, \omega) \mathscr{A}(m, \omega)$$

$$+ \sum_{m=1}^{L-1} e^{-jm\omega d} \chi_b(\tau + md, \omega) \mathscr{A}^*(m, -\omega)$$

where $\chi_b(\tau, \omega)$ refers to a single-bit pulse, and

$$\mathscr{A}(m, \omega) = \sum_{i=0}^{L-1-m} C_i C_{i+m}^* e^{-ji\omega d}$$

If the bit envelopes have durations which are no greater than d, it follows that $\chi_b(\tau, \omega) = 0$ for $|\tau| > d$. Thus, in a given portion of the τ, ω plane, only two displaced χ_b functions will be involved in the calculation of $\chi(\tau, \omega)$. The above is illustrated by Fig. 2.5 which shows the regions of influence of the individual bit functions on the χ function of a four-bit word.

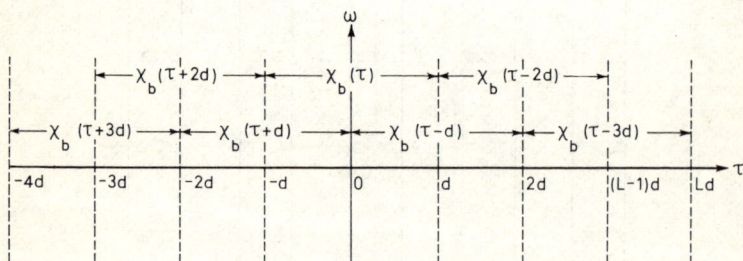

Fig. 2.5. Regions in which the displaced bit χ's exist for a 4-bit word with bit duration $\leqslant d$

It should be particularly noted that along the lines parallel to the ω axis where $\tau = \pm md \{m = 0, 1, \ldots (L - 1)\}$, $\chi_b(\tau - md, \omega)$ is the only non-zero χ_b function.

The function represented by $\mathscr{A}(m, \omega)$ is determined by the code and is independent of the bit shape. For the special case of C_i real (e.g. $0°$, $180°$ phase coding) $\mathscr{A}(m, \omega)$ is equal to $\mathscr{A}^*(m, -\omega)$.

A further discussion of the effects of signal repetition and coding is given in Section 4.3.

2.4 THE EFFECT OF LINEAR FM UPON THE UNCERTAINTY FUNCTION

If linear f.m. is *added* to a complex modulation function $a(t)$, the result is to give a new modulation function of the form

$$a(t) \, e^{jbt^2},$$

i.e. an extra quadratic phase term. Table 2.1 shows that this results in the uncertainty function changing from $\chi(\tau, \omega)$ to

$$e^{-jb\tau^2} \chi[\tau, \omega + 2b\tau].$$

The above result means that the modulus of the new uncertainty function is a sheared version of the modulus of the original uncertainty function. A characteristic which formerly occurred at the co-ordinates τ_1, ω_1 is transferred to

$$\tau = \tau_1$$

$$\omega + 2b\tau = \omega_1$$

That is, to the new co-ordinates $\tau_1, (\omega_1 - 2b\tau_1)$. The effect is illustrated in Fig. 2.6.

When evaluating formulae numerically, some confusion can result over the choice of the parameter b. It is convenient, in algebraic manipulation, to use the single letter b; however, when using the results it is more useful to think in terms of the physical 'frequency sweep'.

Interpreting instantaneous frequency as

$$\omega_i = \frac{d\phi}{dt}$$

the frequency sweep, in time t, caused by the quadratic phase term will be $2bt$. Thus, for an envelope of duration d, the quadratic phase term will lead to a total frequency sweep

$$\Delta = 2bd. \tag{2.29}$$

It follows that, when using results based upon the linear f.m. formulae, b should be given the value

$$b = \frac{\Delta}{2d}. \tag{2.30}$$

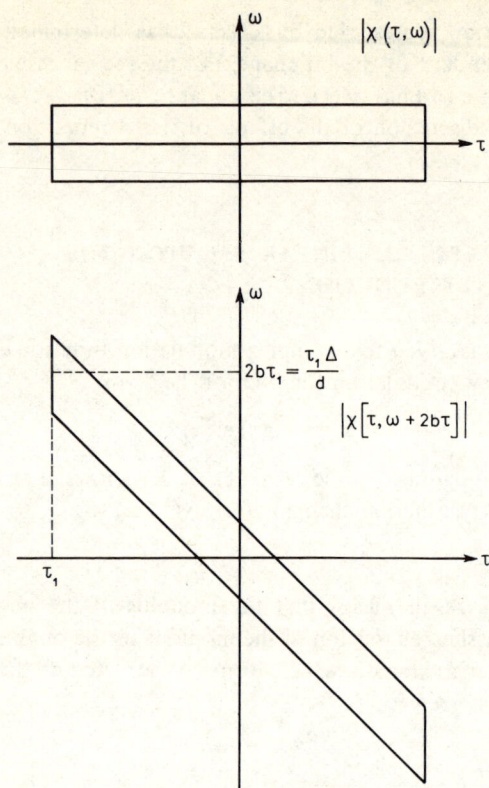

Fig. 2.6. The shearing effect of linear FM upon the area of uncertainty

A term often used in the literature is 'dispersion factor'. Using the above notation, the dispersion factor is equal to the dimensionless product Δd. For the common practical situation of a fixed envelope duration, an increase in the dispersion factor means an increase in the total frequency sweep Δ.

2.5 PROOFS OF THE PROPERTIES OF $\chi(\tau, \omega)$

This section contains the proofs of the mathematical properties of $\chi(\tau, \omega)$ which were stated in Section 2.2.

2.5.1 Proof of Equation 2.15

Equation 2.15 follows, by inspection, from the definitions of $\chi(\tau, \omega)$ and the Fourier transform.

2.5.2 Proof of Equation 2.16

From the results of Section 6.7

$$\mathscr{F}\{a^*(t)\} = A^*(-j\omega).$$

Hence

$$\mathscr{F}\{a^*(t+\tau)\} = e^{j\omega\tau}A^*(-j\omega).$$

Expanding Equation 2.15 by means of the convolution theorem gives

$$\chi(\tau, \omega) = \frac{1}{2\pi}\int\limits_{-\infty}^{\infty} e^{jx\tau}A^*(-jx)A\left[j(\omega-x)\right]dx.$$

Changing the dummy variable to $-x$ gives

$$\chi(\tau, \omega) = \frac{1}{2\pi}\int\limits_{-\infty}^{\infty} A^*(jx)A\left[j(\omega+x)\right]e^{-jx\tau}dx,$$

which is Equation 2.16.

2.5.3 Proof of Equation 2.17

Since

$$[x\,y\,z]^* = x^*y^*z^*,$$

Equation 2.14 leads to

$$\chi^*(\tau, \omega) = \int\limits_{-\infty}^{\infty} a^*(t)a(t+\tau)\,e^{j\omega t}\,dt$$

Changing the dummy variable to $(t-\tau)$ gives

$$\chi^*(\tau, \omega) = \int\limits_{-\infty}^{\infty} a^*(t-\tau)a(t)\,e^{j\omega(t-\tau)}\,dt$$

which becomes

$$\chi^*(\tau, \omega) = e^{-j\omega\tau}\int\limits_{-\infty}^{\infty} a(t)a^*(t-\tau)\,e^{j\omega t}\,dt.$$

The above integral is equal, by inspection, to $\chi(-\tau, -\omega)$, thus proving equation 2.17.

2.5.4 Proof of Equation 2.18

From Equation 2.17

$$|\chi^*(\tau, \omega)| = |\chi(-\tau, -\omega)|.$$

Also, since

$$|\chi^*(\tau, \omega)| = |\chi(\tau, \omega)|,$$

Equation 2.18 follows.

2.5.5 Proof of Equation 2.19

The equality in Equation 2.19 follows by inspection, remembering that $|x|^2 = x\,x^*$.

The inequality in Equation 2.19 can be obtained from the Schwarz inequality [6]. Since

$$\left| \int\limits_{-\infty}^{\infty} f(x)\,g(x)\,\mathrm{d}x \right|^2 \leqslant \int\limits_{-\infty}^{\infty} |f(x)|^2\,\mathrm{d}x \int\limits_{-\infty}^{\infty} |g(x)|^2\,\mathrm{d}x$$

it follows that

$$|\chi(\tau, \omega)|^2 = \left| \int\limits_{-\infty}^{\infty} a(t)\,a^*(t+\tau)\,\mathrm{e}^{-\mathrm{j}\omega t}\,\mathrm{d}t \right|^2 \leqslant \int\limits_{-\infty}^{\infty} |a(t)|^2\,\mathrm{d}t \int\limits_{-\infty}^{\infty} |a^*(t+\tau)|^2\,\mathrm{d}t.$$

The application of Parseval's theorem (Appendix 2) shows that the two right-hand integrals are equal. Since it has already been shown that

$$\int\limits_{-\infty}^{\infty} |a(t)|^2\,\mathrm{d}t = \chi(0, 0)$$

it follows that

$$|\chi(\tau, \omega)|^2 \leqslant [\chi(0, 0)]^2$$

which proves Equation 2.19.

2.5.6 Proof of Equation 2.20

The proof of Equation 2.20 is obtained by replacing $|\chi(\tau, \omega)|^2$ by $\chi(\tau, \omega)\,\chi^*(\tau, \omega)$. $\chi(\tau, \omega)$ is expanded by means of Equation 2.14, and $\chi^*(\tau, \omega)$ by Equation 2.16. The result is

$$\iint\limits_{-\infty}^{\infty} |\chi(\tau, \omega)|^2\,\mathrm{d}\tau\,\mathrm{d}\omega$$

$$= \iiint\limits_{-\infty}^{\infty} a(t)\,a^*(t+\tau)\,\mathrm{e}^{-\mathrm{j}\omega t}\,\mathrm{d}t \;\frac{1}{2\pi} \int\limits_{-\infty}^{\infty} A(\mathrm{j}x)\,A^*[\mathrm{j}(\omega+x)]\;\mathrm{e}^{\mathrm{j}x\tau}\,\mathrm{d}x\,\mathrm{d}\tau\,\mathrm{d}\omega.$$

The right-hand side may be re-arranged to give

$$\frac{1}{2\pi} \iint\limits_{-\infty}^{\infty} a(t)A(jx) \int\limits_{-\infty}^{\infty} a^*(t+\tau)\, e^{jx\tau}\, d\tau \int\limits_{-\infty}^{\infty} A^*[j(\omega+x)]\, e^{-j\omega t}\, d\omega\, dt\, dx$$

Since the integrals over τ and ω, in the above expression, contain all the τ and ω functions they may be evaluated separately. Now

$$\int\limits_{-\infty}^{\infty} a^*(t+\tau)\, e^{jx\tau}\, d\tau = \int\limits_{-\infty}^{\infty} a^*(\tau)\, e^{j(\tau-t)x}\, d\tau = e^{-jxt}\, A^*(jx)$$

Also

$$\int\limits_{-\infty}^{\infty} A^*[j(\omega+x)]\, e^{-j\omega t}\, d\omega = \int\limits_{-\infty}^{\infty} A^*(j\omega)\, e^{-j(\omega-x)t}\, d\omega$$

$$= 2\pi\, e^{jxt} a^*(t).$$

The above results follow from the definitions of the Fourier transform and its inverse. Substitution in the expression for

$$\iint\limits_{-\infty}^{\infty} |\chi(\tau,\omega)|^2\, d\tau\, d\omega$$

gives

$$\iint\limits_{-\infty}^{\infty} |\chi(\tau,\omega)|^2\, d\tau\, d\omega = \int\limits_{-\infty}^{\infty} a(t)a^*(t)\, dt \int\limits_{-\infty}^{\infty} A(jx)A^*(jx)\, dx$$

Parseval's theorem shows the last integral to be equal to 2π times the middle integral. It has already been shown that

$$\int\limits_{-\infty}^{\infty} a(t)a^*(t)\, dt = \int\limits_{-\infty}^{\infty} |a(t)|^2\, dt = \chi(0,0)$$

Hence

$$\frac{1}{2\pi} \iint\limits_{-\infty}^{\infty} |\chi(\tau,\omega)|^2\, d\tau\, d\omega = [\chi(0,0)]^2$$

which is Equation 2.20.

2.5.7 Proof of Equations 2.21 and 2.22

Defining

$$\chi_1(\tau,\omega) = \int\limits_{-\infty}^{\infty} b(t)b^*(t+\tau)\, e^{-j\omega t}\, dt$$

and substituting

$$b(t) = a(t)\, e^{jbt^2},$$

gives

$$\chi_1(\tau, \omega) = \int_{-\infty}^{\infty} a(t)\, e^{jbt^2}\, a^*(t+\tau)\, e^{-j(t+\tau)^2 b}\, e^{-j\omega t}\, dt$$

$$= \int_{-\infty}^{\infty} a(t)a^*(t+\tau)\, e^{-jb\tau^2}\, e^{-j(\omega + 2b\tau)t}\, dt.$$

Hence

$$\chi_1(\tau, \omega) = e^{-jb\tau^2}\, \chi[\tau, \omega + 2b\tau]$$

proving Equation 2.21.
   Similarly taking

$$B(j\omega) = e^{jb\omega^2}\, A(j\omega)$$

gives

$$\chi_1(\tau, \omega) = \frac{1}{2\pi} \int_{-\infty}^{\infty} e^{-jbx^2} A^*(jx)\, e^{j(\omega + x)^2 b} A\,[j(\omega + x)]\, e^{-jx\tau}\, dx$$

$$= \frac{1}{2\pi} \int_{-\infty}^{\infty} A^*(jx)A\,[j(\omega + x)]\, e^{jb\omega^2}\, e^{-j(\tau - 2b\omega)x}\, dx.$$

Hence

$$\chi_1(\tau, \omega) = e^{jb\omega^2}\, \chi[\tau - 2b\omega, \omega]$$

proving Equation 2.22.

2.5.8 Proof of Equations 2.23 and 2.24

If $b(t) = a(\alpha t)$

$$\chi_1(\tau, \omega) = \int_{-\infty}^{\infty} a(\alpha t)a^*\,[(t+\tau)\alpha]\, e^{-j\omega t}\, dt$$

then

$$\chi_1(\tau, \omega) = \frac{1}{|\alpha|} \int_{-\infty}^{\infty} a(t)a^*(t+\alpha\tau)\, e^{-j\omega t/\alpha}\, dt$$

$$= \frac{1}{|\alpha|}\, \chi[\alpha\tau, \omega/\alpha]$$

which is Equation 2.23.

Also, if $B(j\omega) = A(j\alpha\omega)$

then

$$\chi_1(\tau, \omega) = \frac{1}{2\pi} \int\limits_{-\infty}^{\infty} A^*(j\alpha x) A \left[j\alpha(\omega + x)\right] e^{-jx\tau} \, dx$$

changing the dummy variable to x/α gives

$$\chi_1(\tau, \omega) = \frac{1}{2\pi|\alpha|} \int\limits_{-\infty}^{\infty} A^*(jx) A \left[j(\alpha\omega + x)\right] e^{-jx\tau/\alpha} \, dx$$

$$= \frac{1}{|\alpha|} \chi[\tau/\alpha, \alpha\omega]$$

which is Equation 2.24.

2.5.9 Proof of Equation 2.25

To prove Equation 2.25 let

$$b(t) = \sum_{i=0}^{n} C_i a(t - ik)$$

Hence

$$b(t) = C_0 a(t) + C_1 a(t - k) \ldots + C_n a(t - nk)$$
$$b^*(t + \tau) = C_0^* a^*(t + \tau) + C_1^* a^*(t + \tau - k) \ldots + C_n^* a^*(t + \tau - nk)$$

giving

$$b(t)b^*(t + \tau)$$

$$= C_0 C_0^* a(t)a^*(t + \tau) \qquad + C_1 C_0^* a(t - k)a^*(t + \tau) \ldots \qquad +$$
$$+ C_n C_0^* a(t - nk)a^*(t + \tau)$$

$$+ C_0 C_1^* a(t)a^*(t + \tau - k) \quad + C_1 C_1^* a(t - k)a^*(t + \tau - k) \ldots \quad +$$
$$+ C_n C_1^* a(t - nk)a^*(t + \tau - k)$$

$$\vdots \qquad\qquad \vdots \qquad\qquad \vdots$$

$$+ C_0 C_n^* a(t)a^*(t + \tau - nk) + C_1 C_n^* a(t - k)a^*(t + \tau - nk) \ldots +$$
$$+ C_n C_n^* a(t - nk)a^*(t + \tau - nk)$$

Since

$$\int_{-\infty}^{\infty} a(t+x)a^*(t+y)\,e^{-j\omega t}\,dt = \int_{-\infty}^{\infty} a(t)a^*(t-x+y)\,e^{-j\omega(t-x)}\,dt$$

$$= e^{j\omega x}\chi[y-x,\,\omega]$$

it follows that

$$\chi_1(\tau,\,\omega) = \int_{-\infty}^{\infty} b(t)b^*(t+\tau)\,e^{-j\omega t}\,dt$$

$$= C_0 C_0^* \chi(\tau,\,\omega) \qquad\quad + e^{-j\omega k}C_1 C_0^* \chi(\tau+k,\,\omega)\,\ldots \qquad\quad +$$

$$+ e^{-j\omega n k}C_n C_0^* \chi(\tau+nk,\,\omega)$$

$$+ C_0 C_1^* \chi(\tau-k,\,\omega) \;+ e^{-j\omega k}C_1 C_1^* \chi[\tau,\,\omega]\,\ldots \qquad\qquad +$$

$$+ e^{-j\omega n k}C_n C_1^* \chi[\tau+(n-1)k,\,\omega]$$

$$\vdots \qquad\qquad\qquad \vdots \qquad\qquad\qquad \vdots$$

$$+ C_0 C_n^* \chi(\tau-nk,\,\omega) + e^{-j\omega k}C_1 C_n^* \chi[\tau-(n-1)k,\,\omega]\,\ldots +$$

$$+ e^{-j\omega n k}C_n C_n^* \chi(\tau,\,\omega)$$

Hence

$$\chi_1(\tau,\,\omega)$$

$$= e^{-j\omega n k}\chi(\tau+nk,\,\omega)\{C_n C_0^*\} + \ldots$$

$$+ e^{-j\omega k}\chi(\tau+k,\,\omega)\{C_1 C_0^* + C_2 C_1^* \,e^{-j\omega k}\ldots + C_n C_{n-1}^* \,e^{-j\omega(n-1)k}\}$$

$$+ \chi(\tau,\,\omega)\{C_0 C_0^* + C_1 C_1^* \,e^{-j\omega k}\ldots + C_n C_n^* \,e^{-j\omega n k}\}$$

$$+ \chi(\tau-k,\,\omega)\{C_0 C_1^* + C_1 C_2^* \,e^{-j\omega k}\ldots + C_{n-1}C_n^* \,e^{-j\omega(n-1)k}\}$$

$$\cdots + \chi(\tau-nk,\,\omega)\{C_0 C_n^*\}$$

which leads to

$$\chi_1(\tau,\,\omega) = \sum_{m=1}^{n} e^{-j\omega m k}\chi(\tau+mk,\,\omega) \sum_{i=0}^{n-m} C_{i+m}C_i^* e^{-j\omega i k}$$

$$+ \sum_{m=0}^{n} \chi(\tau-mk,\,\omega) \sum_{i=0}^{n-m} C_i C_{i+m}^* \,e^{-j\omega i k}$$

Replacing n by $n-1$ so that the new 'n' is equal to the number of pulses gives

$$\chi_1(\tau,\,\omega) = \sum_{m=1}^{n-1} e^{-j\omega m k}\chi(\tau+mk,\,\omega) \sum_{i=0}^{n-1-m} C_{i+m}C_i^* e^{-j\omega i k}$$

$$+ \sum_{m=0}^{n-1} \chi(\tau-mk,\,\omega) \sum_{i=0}^{n-1-m} C_i C_{i+m}^* \,e^{-j\omega i k}$$

2.5.10 Special cases of Equation 2.25

Three special cases will now be considered.

(1) Gated source.

$$f(t) = \sum_{i=0}^{n-1} |a(t-ik)| \cos[\omega_0 t + \phi(t-ik)]$$

Hence

$$f_a(t) = \sum_{i=0}^{n-1} a(t-ik) e^{j\omega_0 t}$$

which corresponds to $C_i = 1$. Thus the coefficient of $\chi(\tau + mk, \omega)$ is a geometric progression given by

$$e^{-j\omega mk}\{1 + e^{-j\omega k} + \ldots + e^{-j\omega k(n-1-m)}\}$$

$$= e^{-j\omega mk}\left\{\frac{e^{-j\omega k(n-m)} - 1}{e^{-j\omega k} - 1}\right\}$$

Similarly the coefficient of $\chi(\tau - mk, \omega)$ is

$$\left\{\frac{e^{-j\omega k(n-m)} - 1}{e^{-j\omega k} - 1}\right\}$$

By defining $\theta = \dfrac{\omega k}{2}$ and using the identity

$$(1 - e^{-2jx}) = 2j\, e^{-jx} \sin(x)$$

the above coefficients can be written as follows:

$$\chi(\tau + mk, \omega) \rightarrow e^{-j(n-1+m)\theta}\left\{\frac{\sin[(n-m)\theta]}{\sin\theta}\right\}$$

$$\chi(\tau - mk, \omega) \rightarrow e^{-j(n-1-m)\theta}\left\{\frac{\sin[(n-m)\theta]}{\sin\theta}\right\}$$

which is expressed compactly by saying that the coefficient of $\chi(\tau - mk, \omega)$ is

$$e^{-j(n-1-m)\theta}\left\{\frac{\sin[(n-|m|)\theta]}{\sin\theta}\right\}$$

for $0 \leqslant |m| \leqslant n - 1$. The expression for $\chi_1(\tau, \omega)$ then becomes:

$$\sum_{m=-(n-1)}^{(n-1)}\left\{e^{-j(n-1-m)\theta}\left\{\frac{\sin[(n-|m|)\theta]}{\sin\theta}\right\}\chi[\tau - mk, \omega]\right\}$$

(2) Pulsed source.

$$f(t) = \sum_{i=0}^{n-1} |a(t-ik)| \cos[\omega_0(t-ik) + \phi(t-ik)]$$

Hence

$$f_a(t) = \sum_{i=0}^{n-1} e^{-j\omega_0 ki} a(t-ik) e^{j\omega_0 t}$$

which corresponds to $C_i = e^{-j\omega_0 ki}$. Thus

$$C_{i+m} C_i^* = e^{j\omega_0 km}$$

Since the above products are not functions of i, the coefficients of $\chi(\tau + mk, \omega)$ and $\chi(\tau - mk, \omega)$ are once more geometric progresssions. The same reasoning as in the gated case leads to the result

$$\chi_1(\tau, \omega) =$$

$$\sum_{m=-(n-1)}^{n-1} \left\{ e^{j\omega_0 km} e^{-j(n-1-m)\theta} \left\{ \frac{\sin[(n-|m|)\theta]}{\sin\theta} \right\} \chi[\tau-mk, \omega] \right\}$$

(3) Non-coherent source. This is the case for $C_i = e^{j\phi_i}$ where ϕ_i takes on random values between 0 and 2π radians. The co-efficient of $\chi(\tau - mk, \omega)$ for $m = 0$ is a geometric progression

$$\sum_{i=0}^{n-1} C_i C_i^* e^{-j\omega ik} = \sum_{i=0}^{n-1} e^{-j\omega ik}$$

$$= e^{-j(n-1)\theta} \left\{ \frac{\sin[n\theta]}{\sin\theta} \right\}$$

where $\theta = \omega k/2$. The coefficients of $\chi(\tau \pm mk, \omega)$ for $m \pm 0$ are not geometric progressions and cannot be expressed in a closed form.

2.5.11 Proof of Equation 2.26

Equation 2.26 can be proved by using the function $a(t) + b(t)$ in the definition of $\chi(\tau, \omega)$:

$$\chi(\tau, \omega) = \int_{-\infty}^{\infty} [a(t)+b(t)][a^*(t+\tau)+b^*(t+\tau)] e^{-j\omega t} dt$$

Multiplying out gives

$$\chi(\tau, \omega) = \chi_{aa}(\tau, \omega) + \chi_{bb}(\tau, \omega) + \chi_{ab}(\tau, \omega) + \chi_{ba}(\tau, \omega)$$

where

$$\chi_{uv}(\tau, \omega) = \int_{-\infty}^{\infty} u(t)v^*(t+\tau) e^{-j\omega t} dt$$

It follows that

$$\chi_{ba}(\tau, \omega) = \int_{-\infty}^{\infty} b(t)a^*(t+\tau)\, e^{-j\omega t}\, dt$$

Changing the dummy variable to $(t - \tau)$ gives

$$\chi_{ba}(\tau, \omega) = e^{j\omega\tau} \int_{-\infty}^{\infty} a^*(t)b(t-\tau)\, e^{-j\omega t}\, dt$$

Hence

$$\chi_{ba}(\tau, \omega) = e^{j\omega\tau}\chi_{ab}^*(-\tau, -\omega)$$

Substitution of the above in the expression for $\chi(\tau, \omega)$ gives Equation 2.26.

Chapter 3

WORKED UNCERTAINTY FUNCTION EXAMPLES

This chapter is used to illustrate the method of calculating and representing uncertainty functions.

3.1 THE RECTANGULAR PULSE WITH CONSTANT CARRIER FREQUENCY

For the rectangular pulse with constant carrier frequency the baseband modulating function is given by

$$a(t) = \quad \text{(rectangular pulse, height 1, from } 0 \text{ to } d, \ t \to)$$

Hence

$$a^*(t + \tau) = \quad \text{(rectangular pulse, height 1, from } -\tau \text{ to } -\tau+d, \ t \to)$$

Thus, for $0 < \tau < d$

$$a(t)a^*(t + \tau) = \quad \text{(rectangular pulse, height 1, from } 0 \text{ to } -\tau+d, \ t \to)$$

And for $-d < \tau < 0$

$$a(t)a^*(t + \tau) = \quad \text{(rectangular pulse, height 1, from } -\tau \text{ to } d, \ t \to)$$

Finally, for $|\tau| > d$

$$a(t)a^*(t + \tau) = 0$$

From Equation 2.15

$$\chi(\tau, \omega) = \mathscr{F}\{a(t)a^*(t + \tau)\}$$

Hence for $0 < \tau < d$

$$\chi(\tau, \omega) = \frac{1 - e^{-j(d-\tau)\omega}}{j\omega}$$

or

$$\chi(\tau, \omega) = e^{-j(\omega/2)(d-\tau)} \frac{2}{\omega} \sin\left[\frac{\omega}{2}(d-\tau)\right] \qquad (3.1)$$

For $-d < \tau < 0$

$$\chi(\tau, \omega) = e^{j\omega\tau}\left[\frac{1-e^{-j(d+\tau)\omega}}{j\omega}\right]$$

Hence

$$\chi(\tau, \omega) = e^{-j(\omega/2)(d-\tau)} \frac{2}{\omega} \sin\left[\frac{\omega}{2}(d+\tau)\right] \qquad (3.2)$$

Equations 3.1 and 3.2 can be combined to give

$$\left.\begin{aligned}
&\chi(\tau, \omega) = e^{-j(\omega/2)(d-\tau)} \frac{2}{\omega} \sin\left[\frac{\omega}{2}(d-|\tau|)\right], \quad |\tau| < d \\
&\chi(\tau, \omega) = 0, \quad |\tau| > d
\end{aligned}\right\} \qquad (3.3)$$

The cut along the τ axis of the uncertainty function is given by $|\chi(\tau, 0)|$, where

$$\left.\begin{aligned}
&|\chi(\tau, 0)| = d - |\tau|, \quad |\tau| < d \\
&\chi(\tau, 0) = 0, \quad |\tau| > d
\end{aligned}\right\} \qquad (3.4)$$

The cut along the ω axis of the uncertainty function is given by $|\chi(0, \omega)|$, where

$$|\chi(0, \omega)| = d\left|\frac{\sin(\omega d/2)}{(\omega d/2)}\right| \qquad (3.5)$$

Equations 3.4 and 3.5 are illustrated by Figs 3.1 and 3.2. The full 3-dimensional function of Equation 3.3 is illustrated by the computer plot of Fig. 3.3; Fig. 3.4 is a computer plot showing the area of uncertainty formed by the −6 dB contour of $|\chi(\tau, \omega)|$.

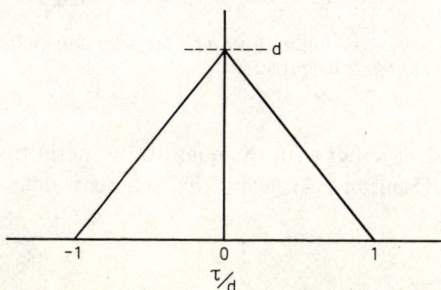

Fig. 3.1. $|\chi(\tau, 0)|$ *for a rectangular pulse with constant carrier frequency*

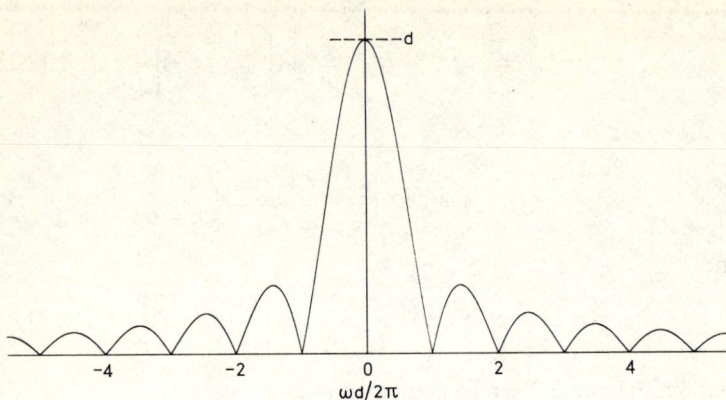

Fig. 3.2. $|\chi(0, \omega)|$ *for a rectangular pulse with constant carrier frequency*

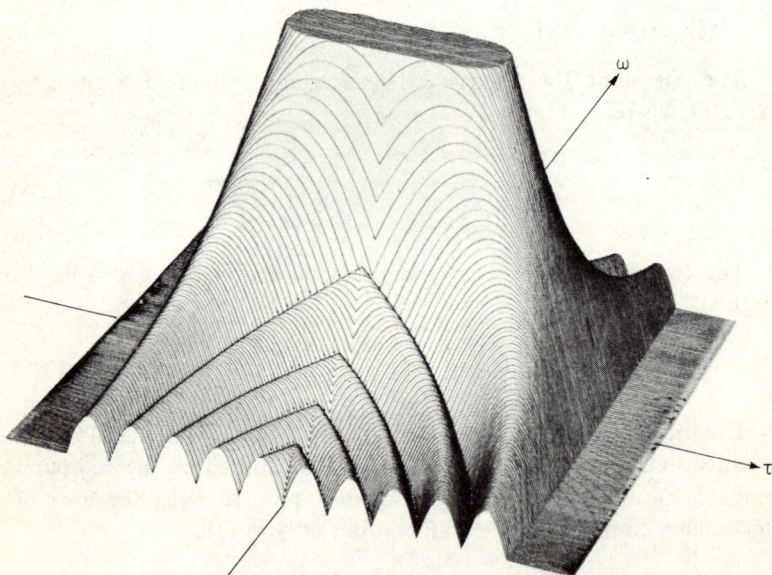

Fig. 3.3. The uncertainty function of a rectangular pulse with constant carrier frequency (truncated at half height)

Using Fig. 3.4 together with the quantitative definitions of precision and resolution (Section 1.4) gives the inherent delay and Doppler precisions as

$$\frac{\tau}{d} = 1 \tag{3.6}$$

$$\frac{\omega d}{2\pi} = 1.2 \tag{3.7}$$

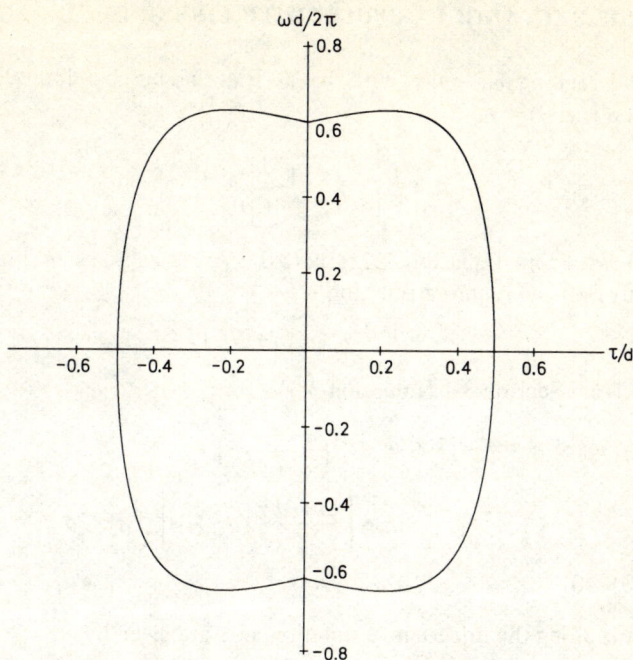

Fig. 3.4. The −6 dB area of uncertainty of a rectangular pulse with constant carrier frequency

while the corresponding figures for resolution are

$$\frac{\tau}{d} = 2 \text{ (for complete resolution)} \qquad (3.8)$$

$$\frac{\omega d}{2\pi} = 637 \text{ (for 60 dB resolution)} \qquad (3.9)$$

The relationship between τ/ω and range/velocity is given in Section 1.4 (Equations 1.6 and 1.7); making the appropriate substitutions leads to the following relationships for inherent range and velocity precision

$$r = 0.5 \, cd \qquad (3.10)$$

$$v = 0.6 \, \frac{c}{f_0 d} \qquad (3.11)$$

Note that $f_0 d$ is equal to the number of RF cycles contained in the pulse.

3.2. THE RECTANGULAR PULSE WITH LINEAR FM

For the rectangular pulse with linear FM the baseband modulating function is given by:

$$a(t) = \left[\underset{0 \qquad d}{\underset{t \longrightarrow}{\rule{0pt}{0pt}}} \right]^{-1} e^{jbt^2}$$

It follows, from Equation 2.21 that, if $\chi_1(\tau, \omega)$ applies to the pulse without FM, the required function is

$$e^{-jb\tau^2} \chi_1 [\tau, \omega + 2b\tau]$$

Hence from Section 3.1 (Equation 3.3)

$$\chi(\tau, \omega) = e^{-jb\tau^2} e^{-j(\omega + 2b\tau)(d - \tau)/2} \left(\frac{2}{\omega + 2b\tau} \right)$$

$$\left. \begin{array}{ll} \qquad \sin \left[\dfrac{\omega + 2b\tau}{2} (d - |\tau|) \right], & |\tau| < d \\[2mm] \chi(\tau, \omega) = 0, & |\tau| > d \end{array} \right\} \quad (3.12)$$

The cuts along the uncertainty function axes are given by

$$\left. \begin{array}{ll} |\chi(\tau, 0)| = d \left| \dfrac{\sin [b\tau(d - |\tau|)]}{b\tau d} \right|, & |\tau| < d \\[3mm] |\chi(\tau, 0)| = 0, & |\tau| > d \end{array} \right\} \quad (3.13)$$

$$|\chi(0, \omega)| = d \left| \frac{\sin(\omega d/2)}{(\omega d/2)} \right| \qquad (3.14)$$

Using the relationship $b = \Delta/2d$ (Equation 2.30), $|\chi(\tau, 0)|$ can be written in the form

$$\left. \begin{array}{ll} |\chi(\tau, 0)| = d \left| \dfrac{\sin \left[\dfrac{\tau \Delta}{2} \left(1 - \left| \dfrac{\tau}{d} \right| \right) \right]}{(\tau \Delta/2)} \right|, & \left| \dfrac{\tau}{d} \right| < 1 \\[4mm] |\chi(\tau, 0)| = 0, & \left| \dfrac{\tau}{d} \right| > 1 \end{array} \right\} \quad (3.15)$$

The argument of the sine function is illustrated by Fig. 3.5. Nulls occur in $|\chi(\tau, 0)|$ whenever the graph of Fig. 3.5 passes through integral multiples of π. It can be seen that whenever the dispersion factor $d\Delta$ is large there will be a large number of nulls—with corresponding peaks between them. For a large dispersion factor the last peak will occur in the vicinity of $|\tau|/d = 1$ and will have an approximate value of $2d/(d\Delta)$. Thus for a dispersion factor of 2000 the last peak in $|\chi(\tau, 0)|$ will be about 60 dB down on the main lobe.

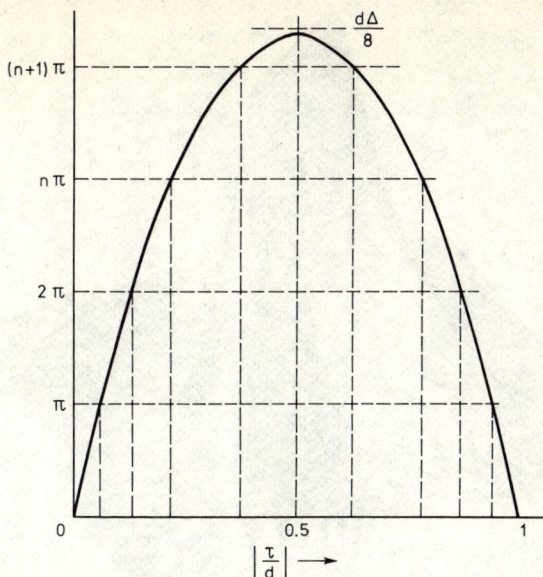

Fig. 3.5. Plot of $\dfrac{\tau\Delta}{2}\left(1 - \left|\dfrac{\tau}{d}\right|\right)$

A further point to note is that for small values of $|\tau|/d$, i.e.

$$\frac{|\tau|}{d} = \frac{|\tau\Delta|}{d\Delta} \leqslant 0.1$$

$|\chi(\tau, 0)|$ can be represented by

$$|\chi(\tau, 0)| \simeq d \left|\frac{\sin(\tau\Delta/2)}{(\tau\Delta/2)}\right|, \quad |\tau\Delta| \leqslant 0.1\, d\Delta \qquad (3.16)$$

Since the $\sin(x)/x$ function has its nulls when x is equal to integral multiples of π, it follows that it is convenient to label the τ and ω axes for this function in terms of

$$\frac{\tau\Delta}{2\pi} \quad \text{and} \quad \frac{\omega d}{2\pi}$$

respectively.

It follows from Equations 3.14 and 3.16 that the Doppler precision and resolution are given by:

$$\frac{\omega d}{2\pi} = 1.2 \text{ (precision)} \qquad (3.17)$$

$$\frac{\omega d}{2\pi} = 637 \text{ (60 dB resolution)} \qquad (3.18)$$

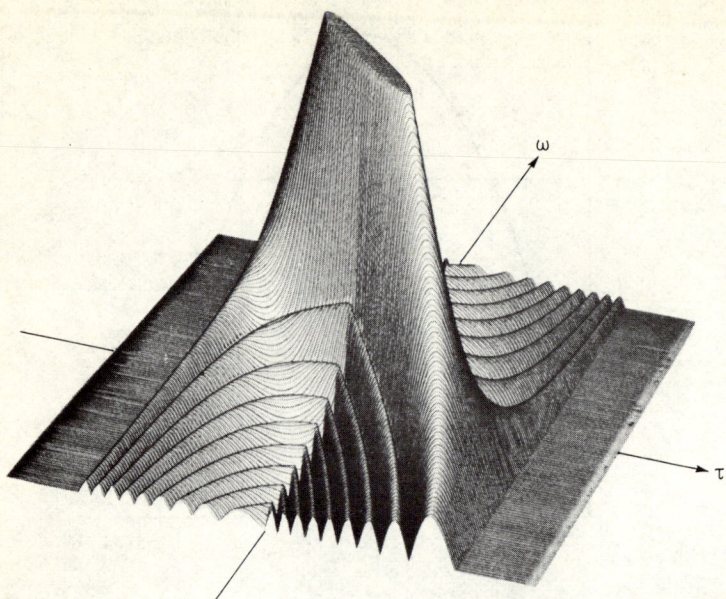

Fig. 3.6. The uncertainty function of a rectangular pulse with linear FM and d∆ = 40 (truncated at half height)

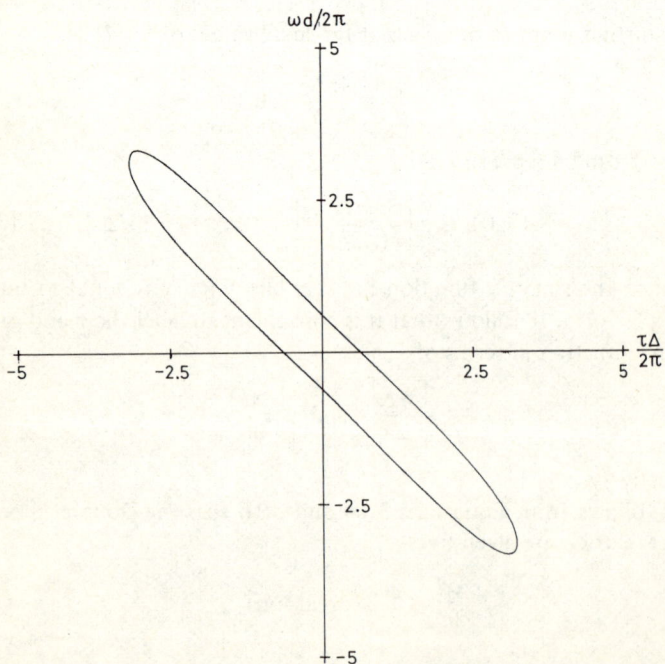

Fig. 3.7. The −6 dB area of uncertainty of a rectangular pulse with linear FM and d∆ = 40

For large values of $d\Delta$ the delay precision will be given by

$$\frac{\tau\Delta}{2\pi} = 1.2 \qquad (3.19)$$

The complete uncertainty function is illustrated by Figs 3.6 and 3.7. Note that there is strong delay–Doppler coupling, the consequences of this are discussed in Sections 4.1, 4.2 and 5.2.

3.3 THE RECTANGULAR PULSE WITH TRIANGULAR FM

The baseband modulating function to be studied in this section is given by

$$c(t) = a(t) + b(t) \qquad (3.20a)$$

where

$$a(t) = \left[\underset{-d \quad \text{o}}{\underbrace{}}^{-1} \right] e^{-jbt^2} \qquad (3.20b)$$
$$t \rightarrow$$

$$b(t) = \left[\underset{\text{o} \quad d}{\underbrace{}}^{-1} \right] e^{jbt^2} \qquad (3.20c)$$
$$t \rightarrow$$

Hence the instantaneous frequency is given by

$$\omega(t) = \underset{-d \quad \text{o} \quad d}{\underbrace{}} \text{-- 2bd}$$
$$t \rightarrow$$

The above definitions maintain the relationship $\Delta = 2bd$; in this case Δ is the total frequency change (rad/s) during either the down sweep or the up sweep.

The results of Section 2.2 (Equations 2.26-2.28) show that the complete uncertainty function is the sum of the uncertainty functions of $a(t)$ and $b(t)$ and of the 'cross uncertainty functions', i.e.

$$\chi(\tau, \omega) = \chi_{aa}(\tau, \omega) + \chi_{bb}(\tau, \omega) + \chi_{ab}(\tau, \omega) + e^{j\omega\tau}\chi_{ab}^*(-\tau, -\omega) \qquad (3.21)$$

where

$$\chi_{uv}(\tau, \omega) = \int_{-\infty}^{\infty} u(t)v^*(t+\tau) e^{-j\omega t} dt \qquad (3.22)$$

If

$$\alpha(t) = \left[\underset{\text{o} \quad d}{\underbrace{}}^{-1} \right] e^{-jbt^2}$$
$$t \rightarrow$$

it follows that $a(t)$ defined by Equation 3.20(b) is given by

$$a(t) = \alpha(t+d)\, e^{j(2t+d)bd} \qquad\qquad (3.23)$$

Equations 3.20(c) and 3.23 allow $\chi_{aa}(\tau, \omega)$ and $\chi_{bb}(\tau, \omega)$ to be found from the results of Section 3.2. The final expressions are

$$
\begin{aligned}
\chi_{aa}(\tau, \omega) &= \exp j\left[\frac{\omega}{2}(\tau+d) - b\tau d\right] \\
&\quad \left(\frac{2}{\omega - 2b\tau}\right)\sin\left[\frac{\omega - 2b\tau}{2}(d-|\tau|)\right], \quad |\tau| < d \\
\chi_{aa}(\tau, \omega) &= 0, \qquad\qquad\qquad\qquad\qquad\qquad\quad |\tau| > d
\end{aligned}
\left.\right\} \quad (3.24)
$$

and

$$
\begin{aligned}
\chi_{bb}(\tau, \omega) &= \exp j\left[\frac{\omega}{2}(\tau-d) - bd\tau\right] \\
&\quad \left(\frac{2}{\omega + 2b\tau}\right)\sin\left[\frac{\omega + 2b\tau}{2}(d-|\tau|)\right], \quad |\tau| < d \\
\chi_{bb}(\tau, \omega) &= 0, \qquad\qquad\qquad\qquad\qquad\qquad\quad |\tau| > d
\end{aligned}
\left.\right\} \quad (3.25)
$$

It is shown in Section 3.3.2 that the cross term $\chi_{ab}(\tau, \omega)$ is given by

$$
\begin{aligned}
\chi_{ab}(\tau, \omega) &= 0, &\tau < 0 \\
\chi_{ab}(\tau, \omega) &= \exp j(\omega\tau/2)\,\exp -j(b\tau^2/2)\,\exp j(\omega^2/8b) \\
&\quad \sqrt{(\pi/4b)}\,\{[C(u_1) + C(u_2)] - j[S(u_1) + S(u_2)]\}, \\
&\qquad\qquad\qquad\qquad\qquad\qquad 0 < \tau < 2d \\
\chi_{ab}(\tau, \omega) &= 0, &\tau > 2d
\end{aligned}
\left.\right\} \quad (3.26)
$$

where $C(u), S(u)$ are Fresnel integrals (see Appendix 3) and

$$
\begin{aligned}
u_1 &= \sqrt{\left(\frac{b\tau^2}{\pi}\right)}\left[1 + \frac{\omega}{2b\tau}\right], &0 < \tau < d \\
u_2 &= \sqrt{\left(\frac{b\tau^2}{\pi}\right)}\left[1 - \frac{\omega}{2b\tau}\right], &0 < \tau < d
\end{aligned}
\left.\right\} \quad (3.27a)
$$

$$
\begin{aligned}
u_1 &= (2d-\tau)\sqrt{\left(\frac{b}{\pi}\right)}\left[1 + \frac{\omega}{2b(2d-\tau)}\right], &d < \tau < 2d \\
u_2 &= (2d-\tau)\sqrt{\left(\frac{b}{\pi}\right)}\left[1 - \frac{\omega}{2b(2d-\tau)}\right], &d < \tau < 2d
\end{aligned}
\left.\right\} \quad (3.27b)
$$

3.3.1 Interpretation of results

The above expressions look very forbidding especially when it is realised that the significant function is the modulus of Equation 3.21. They can, however, be easily interpreted.

$\chi_{aa}(\tau, \omega)$ and $\chi_{bb}(\tau, \omega)$ are sheared functions of the form discussed in Section 3.2; the main part of $\chi_{aa}(\tau, \omega)$ runs between the first and third quadrants as opposed to $\chi_{bb}(\tau, \omega)$ which runs between the second and fourth quadrants. For a large dispersion factor there will be large areas of the τ, ω plane where the two functions do not interact—significant interaction takes place at the origin leading to a large single lobe.

The Fresnel cross functions form a pedestal to the above two functions. It would be very difficult to calculate the exact form of this pedestal but, fortunately, it can be shown to be small compared with the amplitude of $\chi_{aa}(\tau, \omega)$ and $\chi_{bb}(\tau, \omega)$.

For high-precision applications, interest is centred on the performance for moderate values of τ. It can be seen that in this region the effective multiplier of $\chi_{aa}(\tau, \omega)$ and $\chi_{bb}(\tau, \omega)$ is the individual pulse length, d.

Putting $b = \Delta/2d$ and using Equation 3.26 shows that the multiplier for $\chi_{ab}(\tau, \omega)$ is

$$\tfrac{1}{2}d \sqrt{\left/\left(\frac{2\pi}{d\Delta}\right)\right.}.$$

This fact shows that the relative amplitude of $\chi_{ab}(\tau, \omega)$ is small for large values of $d\Delta$. Further information can be obtained from the arguments of the Fresnel integrals.

Substituting for b, Equation 3.27(a) becomes

$$\left. \begin{aligned} u_1 &= \frac{\tau}{d} \sqrt{\left/\left(\frac{d\Delta}{2\pi}\right)\right.} \left[1 + \frac{\omega d}{\tau \Delta}\right], \qquad 0 < \frac{\tau}{d} < 1 \\ u_2 &= \frac{\tau}{d} \sqrt{\left/\left(\frac{d\Delta}{2\pi}\right)\right.} \left[1 - \frac{\omega d}{\tau \Delta}\right], \qquad 0 < \frac{\tau}{d} < 1 \end{aligned} \right\} \qquad (3.28)$$

Hence, along the τ axis

$$u_1 = u_2 = \frac{\tau}{d} \sqrt{\left/\left(\frac{d\Delta}{2\pi}\right)\right.}, \qquad 0 < \frac{\tau}{d} < 1$$

For large values of $d\Delta$, the Fresnel integrals will be equal to 0.5. (Fig. A3.1).

Along the ω axis

$$u_1 = -u_2 = \frac{\omega d}{2\pi} \sqrt{\left/\left(\frac{2\pi}{d\Delta}\right)\right.}$$

For large values of $d\Delta$ the Fresnel integrals will be equal to zero.
Along the main ridges $\omega = \pm \tau \Delta/d$, hence

$$u_1 = \frac{\tau}{d}\sqrt{\left(\frac{d\Delta}{2\pi}\right)}[1 \pm 1], \quad 0 < \frac{\tau}{d} < 1$$

$$u_2 = \frac{\tau}{d}\sqrt{\left(\frac{d\Delta}{2\pi}\right)}[1 \mp 1], \quad 0 < \frac{\tau}{d} < 1$$

leading to Fresnel integrals equal to 0 and (for high $d\Delta$) 0.5.

It should be noted that use of the triangular FM waveform allows target range changes to be distinguished from velocity changes. If either a matched filter or Fourier transform receiver (see Chapter 4) is used to track a given target, the correct setting of the 'tuning' control will be where the output waveform changes from a double to a single pulse.

3.3.2 Proof of Equation 3.26

$\chi_{ab}(\tau, \omega)$ can be calculated in the following manner: from Equation 3.22, after a change of dummy variable

$$\chi_{ab}(\tau, \omega) = e^{j(\omega\tau/2)}\int_{-\infty}^{\infty} a\left(t - \frac{\tau}{2}\right)b^*\left(t + \frac{\tau}{2}\right)e^{-j\omega t}\, dt$$

Using Equations 3.20(b) and 3.20(c)

$$\left|a\left(t - \frac{\tau}{2}\right)b^*\left(t + \frac{\tau}{2}\right)\right| = \underset{-\frac{\tau}{2} \quad \frac{\tau}{2}}{\boxed{}}^{1}, \quad 0 < \tau < d$$

$$\left|a\left(t - \frac{\tau}{2}\right)b^*\left(t + \frac{\tau}{2}\right)\right| = \underset{-d+\frac{\tau}{2} \quad d-\frac{\tau}{2}}{\boxed{}}^{1}, \quad d < \tau < 2d$$

$$\left|a\left(t - \frac{\tau}{2}\right)b^*\left(t + \frac{\tau}{2}\right)\right| = 0, \quad \tau > 2d, \quad \tau < 0$$

Also,

$$\text{Arg}\left\{a\left(t - \frac{\tau}{2}\right)b^*\left(t + \frac{\tau}{2}\right)\right\} = -2bt^2 - \tfrac{1}{2}b\tau^2$$

Equations 3.26 and 3.27 follow when the above information is used with the results of Appendix 3 (Equation A3.6).

Chapter 4

SIGNAL PROCESSING METHODS

This chapter illustrates radar signal processing by describing two 'optimum' receivers—the matched filter receiver and the Fourier transform receiver. The former is basically a range-measuring system which also obtains Doppler information, while the latter can be regarded as a Doppler measuring system which also obtains range information.

The above receivers are optimum, from the precision point of view, in the sense that the target parameters are determined from waveforms which are cuts through the uncertainty function of the transmitted signal. The discussion in Section 1.4 shows that such receivers are not necessarily optimum from the point of view of resolution.

In practice resolution can be improved (at the expense of precision) by using modifications of the 'optimum' receivers. Most practical radar processing methods can be identified as such modifications.

4.1 THE MATCHED FILTER RECEIVER

The matched filter corresponding to a given real signal $f(t)$ can be defined as the filter with an impulse response of $f(-t)$. It is shown in Section 4.4 that:

(1) A matched filter will process $f(t)$ to give an output having a peak value which is greater than (or as great as) the peak output from any other filter having an impulse response of the same energy as the matched filter.
(2) If $f(t)$ is in white noise, the filter matched to $f(t)$ will give the optimum output peak-signal to noise ratio.
(3) A filter having properties (1) and (2) is physically realisable provided that $f(t)$ is of finite duration (i.e. time-limited).

The matched filter receiver will be defined as a superhet receiver using a matched filter in its IF section. A block diagram of such a receiver is shown in Fig. 4.1.

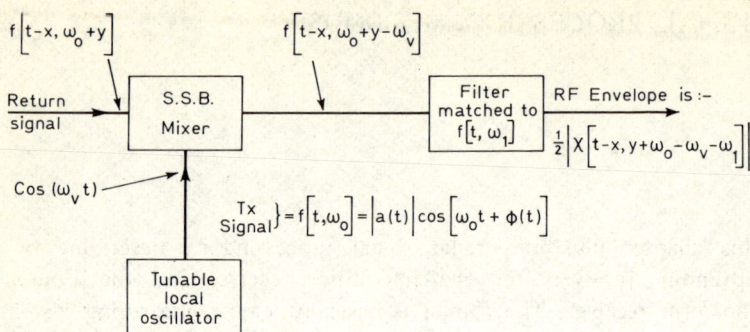

Fig. 4.1. The matched filter receiver

It is shown in Section 4.5 that the output RF envelope of the matched filter receiver is given by

$$\tfrac{1}{2}|X[(t-x), (\omega_0 - \omega_1 - \omega_v + y)]|$$

where $|\chi(\tau, \omega)|$ is the uncertainty function corresponding to

$$|a(t)|\, e^{j\phi(t)}$$

It is assumed that $(\omega_0 - \omega_v)$ is high enough for the results of Section 7.4 to apply.

Thus the RF envelope of the matched filter receiver output is a time function which has the same shape as a cut (taken parallel to the τ axis) through the uncertainty function centred on the x, y co-ordinates of a particular target. The constant value of ω for which the cut is taken can be adjusted by varying the local oscillator frequency ω_v. The effects of variations in ω_v and Doppler shift y are shown in Fig. 4.2; a typical output signal is shown in Fig. 4.3.

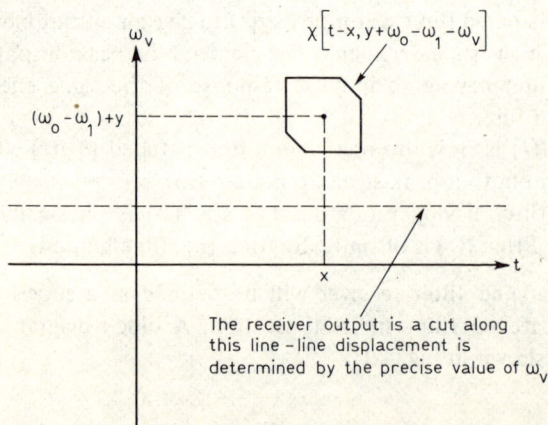

The receiver output is a cut along this line – line displacement is determined by the precise value of ω_v

Fig. 4.2. How ω_v and y determine the uncertainty function cut

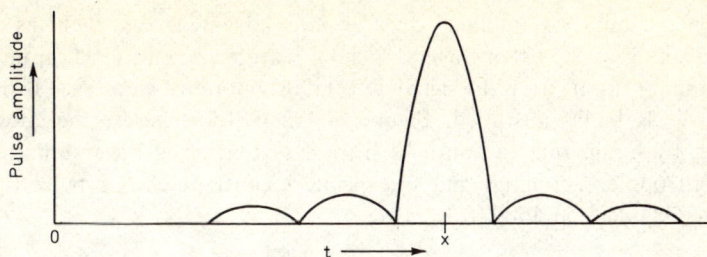

Fig. 4.3. A typical receiver output pulse

Bearing in mind that 'x' represents the time at which a target return is received, Fig. 4.3 appears to indicate that the receiver output pulse starts before the return signal arrives! This paradox is resolved in Section 4.4 where it is shown that the strict mathematical matched filter—being non-causal—cannot be realised. The practical matched filter must include a finite amount of delay—this delay does not affect its optimum properties.

Since the maximum value of the uncertainty function occurs at the origin of the τ, ω plane, the receiver would be 'tuned' to a particular target by adjusting the local oscillator frequency to maximise the echo from that target. Once the receiver was 'tuned', the target delay parameter could be obtained (to within an accuracy set by some threshold) by noting the time difference between the transmitted signal and the peak of the output pulse. The target Doppler shift could be deduced from a knowledge of the value of local oscillator frequency ω_v used to obtain the optimum pulse shape. The threshold in the Doppler measurement would be set by the ability to recognise the optimum pulse shape.

An elaboration of the receiver would be to include the local oscillator in a control loop which would keep the receiver tuned to a particular target once it was selected and acquired. Such a system would give a fast readout of variations in target position (the shift of the 'blip' on the face of an oscilloscope, for example), and a slower readout of variations in target velocity—the speed of the velocity readout being governed by that of the local oscillator control loop. It would be possible to obtain Doppler information faster by using a bank of filters and appropriate switching.

Certain waveforms, a rectangular pulse with linear FM, for example exhibit a coupling between delay and Doppler. This means that if the above tuning procedure were carried out, the pulse shape and amplitude would not change appreciably, the predominant effect of the tuning procedure would be to give a shift in the pulse position. This effect merely conveys, in practical terms, the precision information contained in the uncertainty function of the particular transmitted waveform.

Delay-Doppler coupling can be an advantage as well as a disadvantage. A system using such a waveform will yield a near optimum output, from the signal detectability point of view, even if the receiver is badly mistuned. Figures 4.4 and 4.5 illustrate the above point by comparing the outputs from a system using linear FM (i.e. delay-Doppler coupling) and one using a constant carrier pulse (no delay-Doppler coupling).

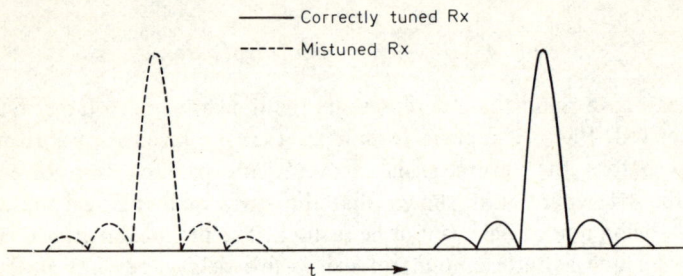

Fig. 4.4. Output of a linear FM matched filter Rx

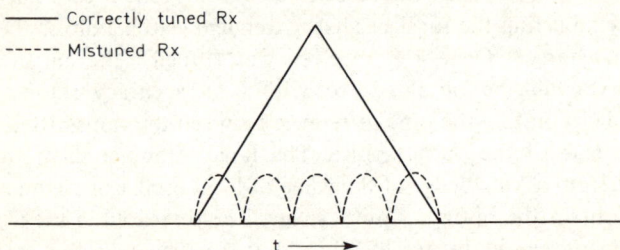

Fig. 4.5. Output of a constant frequency pulse matched filter Rx

All practical radar systems transmit trains of pulses, rather than a single pulse. It was shown in Section 2.3 that the effect of a train of pulses on the uncertainty function was to cause delay ambiguities and to give an improvement in Doppler precision.

4.1.1 A matched filter for repetitive pulses

It should be obvious from the description of the matched filter receiver that pulse trains will always lead to delay ambiguities, e.g. 'second-time-round' ambiguity when the signal delay is greater than the period of the pulse repetition frequency. Improvement in Doppler precision, however, will only be obtained if the filter in the receiver is matched to the train of pulses rather than to a single pulse.

One method of forming a pulse train with inherently high Doppler precision is to transmit an exact repetition of a basic waveform at intervals of k seconds. Figure 4.6 shows a method of converting a filter, $G(j\omega)$, which is matched to a single pulse into a filter matched to such a repetitive train.

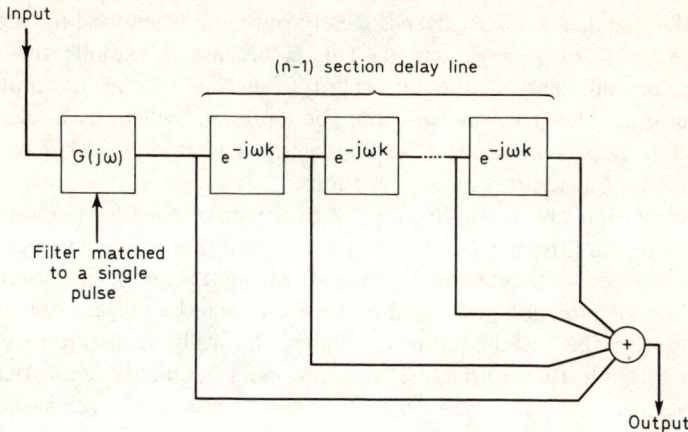

Fig. 4.6. *A matched filter for* n *identical pulses*

A repetitive train of n pulses of the form $f(t)$ can be described by

$$f_R(t) = f(t) + f(t-k) + \ldots + f[t-(n-1)k]$$

Hence the spectrum of the train is given by

$$F_R(j\omega) = F(j\omega)[1 + e^{-j\omega k} + \ldots + e^{-j(n-1)\omega k}]$$

The transfer function of the required matched filter is $F_R^*(j\omega)$, i.e.

$$[F^*(j\omega)][1 + e^{j\omega k} + \ldots + e^{j(n-1)\omega k}]$$

In the above expression $F^*(j\omega)$ is the transfer function of a filter matched to a single pulse (shown as $G(j\omega)$ in Fig. 4.6) while $e^{j\omega k}$ is the transfer function of a network which advances a signal in time by k seconds. Since the advance network is not physically realisable the filter is not realisable. However, a delay at the matched filter output will not affect its optimum properties, hence an equivalent filter which is realisable would have a transfer function

$$F^*(j\omega)[1 + e^{j\omega k} + \ldots + e^{j(n-1)\omega k}]\, e^{-j(n-1)\omega k}$$

That is

$$F^*(j\omega)[1 + e^{-j\omega k} + \ldots + e^{-j(n-1)\omega k}]$$

leading to the scheme of Fig. 4.6.

It can be clearly seen from Fig. 4.6 that the improvement in Doppler precision predicted by Section 2.2 (Equation 2.25) is obtained, physically, by the interaction of the n pulses with each other. Thus although an output will be obtained from the filter of Fig. 4.6 after the first pulse of the train, the system will not achieve its full precision until a time of nk seconds has elapsed.

The matched filter receiver is clearly optimum when used to receive the echo from a single target. This is because it exploits the full precision inherent in the uncertainty function of the transmitted waveform. The receiver will not be optimum when used against multiple targets and clutter, owing to the slow rate of fall-off of the sidelobes of many uncertainty functions.

Resolution can be improved (at the expense of precision) by using a non-matched filter in the receiver—the matched filter characteristics can then be used as a reference against which improvement in resolution and loss of precision and signal-to-noise ratio can be judged. For some waveforms the matched filter although physically realisable may be very difficult to manufacture; in these cases a suitable non-matched filter might be used even when it is only necessary to achieve high precision.

Examples of such alternative schemes are given in Chapter 5.

4.2 THE FOURIER TRANSFORM RECEIVER

The Fourier transform receiver will be defined as a system which calculates the spectrum of the product of the returned signal and an offset version of the transmitted signal. A block diagram of such a receiver is shown in Fig. 4.7.

The Fourier transform receiver is basically a Doppler measuring system which obtains range information as a 'bonus'; in this sense it is the opposite of the matched filter receiver. It is shown in Section 4.7 that the multiplication performed in the Fourier transform receiver leads to a pair of displaced uncertainty functions for each target. The receiver output waveform has the same shape as a cut taken parallel to the ω axis through the modulus of the uncertainty functions; the situation is illustrated by Fig. 4.8.

The amount of separation between the two uncertainty functions resulting from a given target is determined by the frequency of the fixed offset oscillator in Fig. 4.7. If the fixed offset oscillator was omitted (or if its frequency ω_L was not high enough), the side lobes from the two uncertainty functions would overlap and lead to a loss of precision. On the other hand, if ω_L is made sufficiently high the effect of the side lobes of the lower uncertainty function (Fig. 4.8) would be

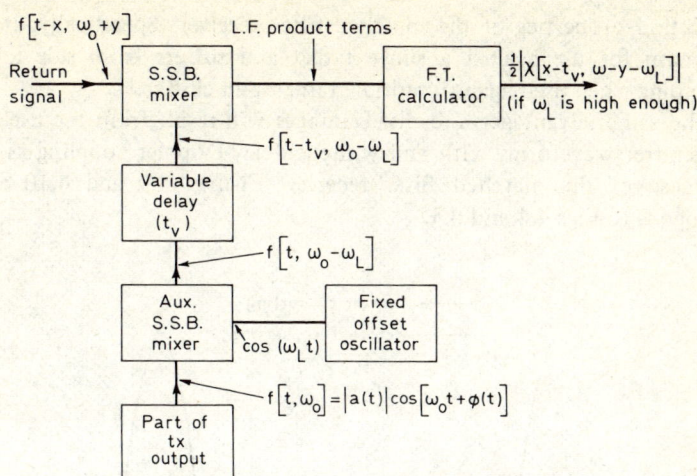

Fig. 4.7. The Fourier transform receiver

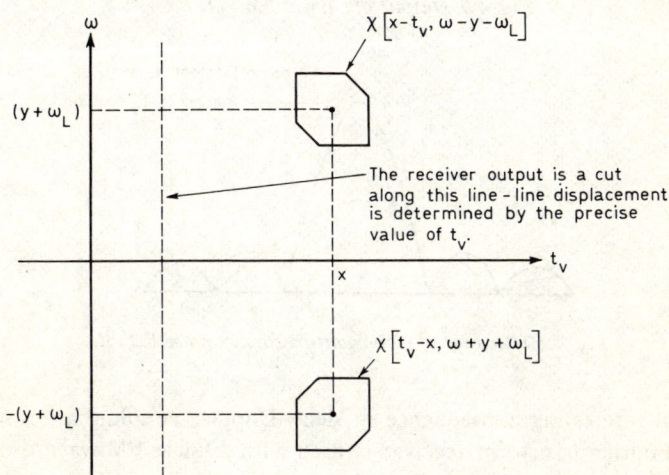

Fig. 4.8. The effect of ω_L and t_v on the receiver output

negligible at the +ve frequencies where the upper uncertainty function is strongest.

To operate the receiver one would adjust the variable delay control to obtain a maximum output pulse amplitude and then read the Doppler offset by noting the frequency at which the peak occurs. If the variable delay control was calibrated the target range could be found by realising that the pulse shape would be optimum when $t_v = x$.

It can be seen that, since the Fourier transform receiver takes a cut through the uncertainty function, it possesses all the precision and resolution properties of the matched filter receiver. Specifically, it is optimum for use against a single target and suffers from side lobe swamping when used against multiple targets and clutter.

The same advantages and disadvantages will result from the use of transmitter waveforms with and without delay-Doppler coupling as in the case of the matched filter receiver. Figures 4.9 and 4.10 are analogous to Figs 4.4 and 4.5.

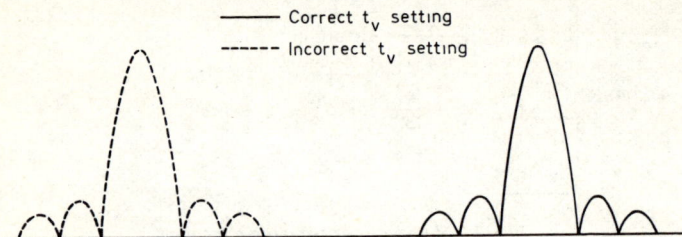

Fig. 4.9. Output of a linear FM F.T. Rx

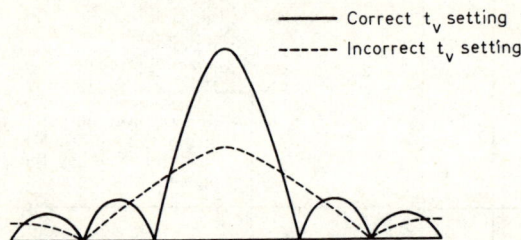

Fig. 4.10. Output of a constant frequency pulse F.T. Rx

An interesting consequence of delay-Doppler coupling arises when the Fourier transform receiver is used with a linear FM waveform. In this case there is often no need to use either the fixed offset oscillator or the variable delay line. The requirements are

(1) The target must be sufficiently far from the transmitter.
(2) The dispersion factor $d\Delta$ must be large.

The above situation arises because the uncertainty function of a linear FM signal with a high dispersion factor is in the form of a long ridge inclined at an angle to the τ axis, (see Section 3.2), and provides its own offset if the range delay x is large enough. The effect is illustrated by Fig. 4.11 which is drawn for y and ω_L equal to zero. The fact that the above situation makes it impossible to distinguish between changing target range and changing velocity is a consequence of the

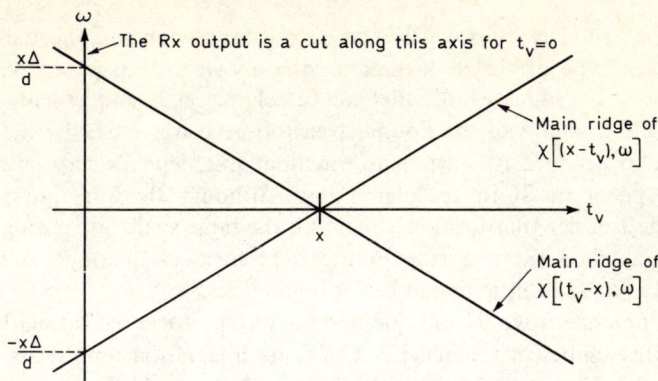

Fig. 4.11. Uncertainty function situation resulting from the use of linear FM in a F.T. Rx with no offset frequency or variable delay

transmitted waveform rather than of the receiver. This effect is discussed further in Section 5.2.

As discussed in Sections 2.3 and 4.1, one of the effects of transmitting a train of pulses is to improve the inherent Doppler precision by concentrating the uncertainty function about multiples of the pulse repetition frequency. In a matched filter receiver additional circuitry is required to exploit this improvement (see Fig. 4.6). The effective number of pulses (using a matched filter receiver) is determined by the number of delay line sections and a time of nk seconds must elapse before the full precision is achieved.

The precision improvement due to repetition is obtained automatically with the Fourier transform receiver—the spectrum of a train of pulses is in the form of narrow spikes and reduces to a 'line spectrum' for an infinite number of pulses. The same time penalty has to be paid for the improvement as in the case of the matched filter receiver. A true Fourier transform calculator would carry out its integration over an infinitely long period, a practical calculator would have a finite integration time and it would be this time which would determine the effective number of pulses and when the full result was available.

Specifically, a Fourier transform calculator which processed mixer output batches of duration nk seconds according to the formula

$$\int_0^{nk} f(t)\, e^{-j\omega t}\, dt$$

would result in a Fourier transform receiver having the same precision as a matched filter receiver using a filter matched to n pulses.

A theoretical advantage of the Fourier transform receiver is its suitability for the processing of a large class of signals as opposed to the matched filter receiver which requires a different and complicated filter for each type of signal. In practice the above advantage can only be exploited if a suitable multiplier and calculator can be implemented.

A disadvantage of the Fourier transform receiver might be the time taken to produce its output information; this depends very much on the type of transform calculator used. Although the integration time for the Fourier transform calculation is the same as the processing time of a matched filter, a true Fourier transform calculation has to be carried out at an infinite number of frequencies.

In practice it would only be necessary to perform the calculation at a finite number of frequencies. The finite integration time sets a limit to the maximum achievable precision, so there would be no advantage in spacing the calculation frequencies much closer than the 'precision frequency'. Also a knowledge of the likely spread of target parameters would allow the calculation to be performed over a finite range.

It follows from the above that, although the processing time of a practical Fourier transform receiver would always be slightly greater than that of a comparable matched filter receiver, whether it would be very much greater depends upon whether the calculation is carried out in serial or parallel form. Practical Fourier transform schemes are discussed briefly in Section 4.6.

4.2.1 Improvement of resolution by weighting

As in the case of the matched filter receiver, the resolution performance of the Fourier transform receiver may not be optimum. Resolution in the matched filter receiver can be improved by frequency weighting (i.e. the use of a non-matched filter)—the corresponding operation in the Fourier transform receiver is time-weighting of the product term prior to calculating its Fourier transform.

The effective action of unweighted Fourier transform or matched filter receivers is to replace each point target 'spike' by the 3-dimensional shape of the uncertainty function. Weighting has the effect of replacing each spike by a shape other than the uncertainty function—hopefully a shape with improved resolution and not too badly degraded precision properties. A question which arises is: can a Fourier transform receiver time-weighting function lead to the same shape as a given mismatched filter? The answer is yes—theoretically—provided that the time origin of the time-weighting function is set by the variable delay control t_v. In many cases it turns out that the required time-weighting function is not real and so cannot be implemented.

It is shown in Section 4.7 (expression 4.29) that the shape generated by a matched filter receiver using a filter with a corresponding complex analytic impulse response of

$$a*(-t)c(t) e^{j\omega_1 t}$$

is given by

$$\frac{1}{2} \left| \int_{-\infty}^{\infty} a(t-x-z)a*(-z)c(z) e^{-j(\omega_0 - \omega_1 + y - \omega_v)z} dz \right| \quad (4.1)$$

Also the shape generated by a Fourier transform receiver using a multiplying function of $m(t - t_v)$ just prior to calculating the Fourier transform is

$$\frac{1}{2} \left| \int_{-\infty}^{\infty} a(t_v - x - z)a*(-z)m(-z) e^{-j(\omega_L + y - \omega)z} dz \right| \quad (4.2)$$

Comparison of expressions 4.1 and 4.2 shows that the corresponding variables in the two receivers are t, ω_v and t_v, ω respectively, as expected.

The time weighting function required to produce the same shape as a filter having a corresponding complex analytic impulse response of $b(t) e^{j\omega_1 t}$ is $m(t - t_v)$ where

$$m(t) = \frac{b(-t)}{a*(t)} \quad (4.3)$$

A special case of a time weighting scheme which does not depend upon the above result is discussed in Section 5.2.1.

4.3 THE EFFECT OF FINITE PROCESSING TIME

The uncertainty function of a train of n identical pulses can be obtained from

$$\chi_1(\tau, \omega) = \int_{-\infty}^{\infty} \sum_{i=0}^{n-1} a(t-ik) \sum_{i=0}^{n-1} a*(t+\tau-ik) e^{-j\omega t} dt$$

Although the integration time runs from $-\infty$ to $+\infty$ reference to Fig. 4.12 shows that in the dominant part of the τ, ω plane (i.e. $-k < \tau < k$) the effective integration limits are

$$\int_{0}^{nk-\tau} \text{ for } 0 < \tau < k \quad \text{and} \quad \int_{-\tau}^{nk} \text{ for } -k < \tau < 0.$$

In a practical radar system the signal processing is carried out over the duration of a number of pulses, n, which is very much smaller than the number actually transmitted. Further the processing time is not affected by the delay τ of the incoming signal.

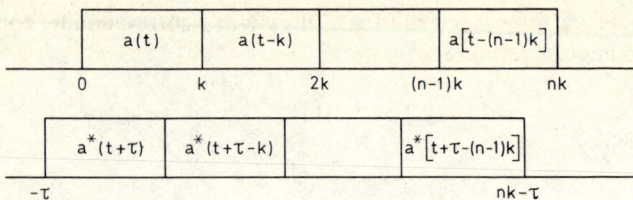

Fig. 4.12. *Product involved in uncertainty function calculation for* $0 < \tau < k$

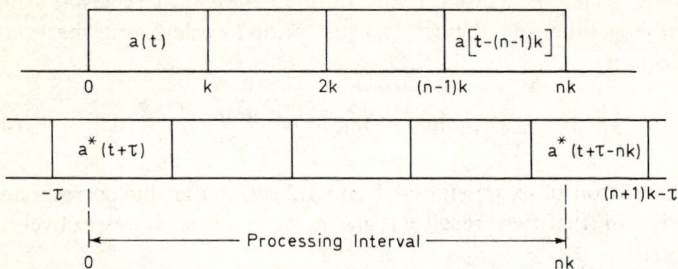

Fig. 4.13. *Situation existing in a practical receiver when* $0 < \tau < k$

A comparison between Figs 4.12 and 4.13 shows that a practical matched filter or Fourier transform receiver will produce an output function $\psi(\tau, \omega)$ which is equal to $\chi_1(\tau, \omega)$ plus an extra integration from $(nk - \tau)$ to nk, i.e.

$$\psi(\tau, \omega) = \chi_1(\tau, \omega) + \int_{-\infty}^{\infty} a[t - (n-1)k]\, a^*[t + \tau - nk]\, e^{-j\omega t}\, dt$$

valid for $0 < \tau < k$.

If $a(t)$ has a duration d, the additional integral is zero for $0 < \tau < k - d$, resulting in the reduction of $\psi(\tau, \omega)$ to $\chi_1(\tau, \omega)$ for that interval. Thus, for low duty cycle pulse trains, the finite processing time has no effect over the dominant part of the τ, ω plane as long as n is put equal to the (integral) number of pulses contained in the signal processing time.

It can also be seen that $\psi(\tau, \omega)$ will not differ appreciably from $\chi_1(\tau, \omega)$, whatever the form of $a(t)$, if n is sufficiently high for the contribution of the additional integral to be insignificant compared with that of $\chi_1(\tau, \omega)$.

$\psi(\tau, \omega)$ can be expected to be significantly different from $\chi_1(\tau, \omega)$ for the case of high duty cycle and n small; this will now be investigated.

By simplification of the additional integral, $\psi(\tau, \omega)$ can be expressed in the following form

$$\psi(\tau, \omega) = \chi_1(\tau, \omega) + e^{-j2(n-1)\theta} \chi(\tau - k, \omega), \qquad 0 < \tau < k$$

Also

$$\psi(\tau, \omega) = \chi_1(\tau, \omega) + \chi(\tau + k, \omega), \quad -k < \tau < 0$$

where $\chi_1(\tau, \omega)$ applies to a train of n pulses, $\chi(\tau, \omega)$ applies to a single pulse and $\theta = \frac{1}{2}\omega k$.

Using the results of Section 2.3, and remembering that $\chi(\tau, \omega) = 0$ for $|\tau| > k$, $\chi_1(\tau, \omega)$ can be written in the form

$$\chi_1(\tau, \omega) = e^{-jn\theta} \frac{\sin[(n-1)\theta]}{\sin \theta} \chi(\tau + k, \omega)$$

$$+ e^{-j(n-1)\theta} \frac{\sin n\theta}{\sin \theta} \chi(\tau, \omega) + e^{-j(n-2)\theta} \frac{\sin[(n-1)\theta]}{\sin \theta}$$

$$\times \chi(\tau - k, \omega) \tag{4.4}$$

valid for $-k < \tau < k$. Substituting Equation 4.4 into the formulae for $\psi(\tau, \omega)$ leads to the expressions

$$\psi(\tau, \omega) = e^{-j(n-1)\theta} \frac{\sin n\theta}{\sin \theta} \{\chi(\tau, \omega) + \chi(\tau - k, \omega)\} \tag{4.5}$$

for $0 < \tau < k$. Also

$$\psi(\tau, \omega) = e^{-j(n-1)\theta} \frac{\sin n\theta}{\sin \theta} \{\chi(\tau, \omega) + \chi(\tau + k, \omega)\} \tag{4.6}$$

for $-k < \tau < 0$.

The above expressions for $\psi(\tau, \omega)$ exhibit a Doppler bar effect as might be expected and also verify the previous statements regarding pulse trains with low duty cycle or large n.

For high duty cycle pulse trains the effect of overlap between $\chi(\tau, \omega)$ and $\chi(\tau \pm k, \omega)$ has to be considered; the calculation of $\psi(\tau, \omega)$ is more straightforward than that of $\chi_1(\tau, \omega)$ as the multipliers of the χ functions are the same.

For waveforms such as linear FM with a high dispersion factor (Section 3.2), $\chi(\tau \pm k, \omega)$ is very small for $|\tau| < k/2$ and both $\chi_1(\tau, \omega)$ and $\psi(\tau, \omega)$ reduce to

$$e^{-j(n-1)\theta} \frac{\sin n\theta}{\sin \theta} \chi(\tau, \omega), \quad -\frac{k}{2} < \tau < \frac{k}{2}$$

4.3.1 Coded waveforms

$\psi(\tau, \omega)$ is significantly different to $\chi_1(\tau, \omega)$ when the transmitted signal consists of contiguous coded words. It is shown below that substitution of the expression given in Section 2.3.1 for $\chi(\tau, \omega)$, in Equations 4.5 and 4.6 leads to

$$\psi(\tau, \omega) = e^{-j(n-1)\theta}\, \frac{\sin n\theta}{\sin \theta} \left\{ \sum_{m=0}^{L-1} \chi_b(\tau - md, \omega) A(m, \omega) \right.$$

$$\left. + \chi_b(\tau - Ld, \omega) A(0, \omega) \right\}, \quad 0 < \tau < Ld \qquad (4.7)$$

Also

$$\psi(\tau, \omega) = e^{-j(n-1)\theta}\, \frac{\sin n\theta}{\sin \theta} \left\{ \chi_b(\tau, \omega) A(0, \omega) \right.$$

$$\left. + \sum_{m=0}^{L-1} \chi_b(\tau + (L-m)d, \omega)\, A(m, \omega) \right\}, \quad -Ld < \tau < 0 \qquad (4.8)$$

where $\chi_b(\tau, \omega)$ refers to a single bit pulse and

$$A(m, \omega) = \sum_{i=0}^{L-1} C_i C_{i+m}^{*}\, e^{-ji\omega d}$$

It will be noted that the function $A(m, \omega)$ differs from $\mathscr{A}(m, \omega)$ of Section 2.3.1 in that the summation extends over the full L bits rather than over $L - m$ bits.

The significance of the above results can be seen by comparing $\psi(\tau, \omega)$ with $\chi_1(\tau, \omega)$ for a $0°$, $180°$ phase-coded waveform. Typical C_i for a 15-bit maximum length shift register code are

i	0	1	2	3	4	5	6	7	8	9	10	11	12	13	14
C_i	−1	1	1	−1	1	1	1	−1	−1	−1	−1	1	−1	1	−1

If the functions are evaluated along the τ axis at integral multiples of the bit length, the results will be independent of the bit pulse shape (i.e. only one displaced bit χ function will be involved at each point and it will have unit amplitude).

$\psi(md, 0)$ is illustrated by Fig. 4.14. It exhibits the two-level effect, characteristic of maximum length codes. The value of n affects only the amplitude and not the shape.

$\chi_1(md, 0)$ is illustrated by Figs 4.15 and 4.16 for $n = 1$ and 10. The desirable two-level effect is absent but it can be seen that the $n = 10$ result is better in this respect than that for $n = 1$. It should be noted that, when n is small, the two-level effect exhibited by $\psi(\tau, \omega)$ is only present when processing extends over an integral number of words. If n is a non-integer, the result will lie somewhere between the extremes illustrated by Figs 4.14 and 4.15. As n becomes larger, the necessity for it to have an integral value becomes less.

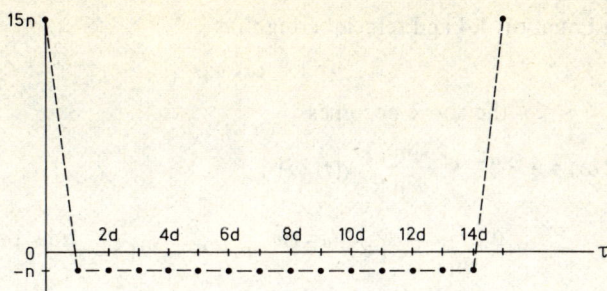

Fig. 4.14. ψ (md, o) for 15-bit maximum length $0°$, $180°$ phase code

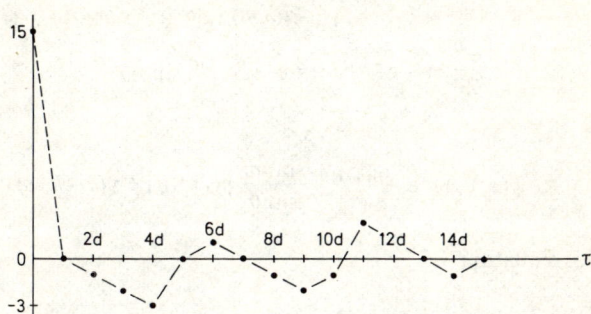

Fig. 4.15. χ_1(md, o) for a 15-bit maximum length $0°$, $180°$ phase code; n = 1

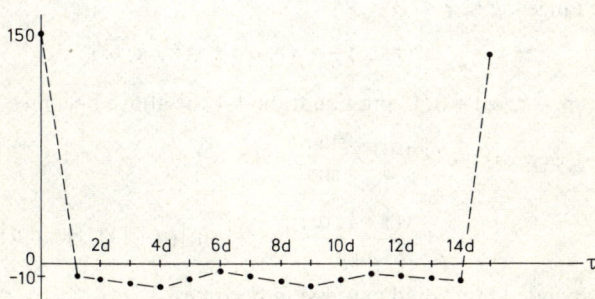

Fig. 4.16. χ_1(md, o) for a 15-bit maximum length $0°$, $180°$ phase code but n = 10

4.3.2 Proof of Equation 4.5

It was shown above that in the range $0 < \tau < k$

$$\psi(\tau, \omega) = \chi_1(\tau, \omega) + e^{-j2(n-1)\theta}\chi(\tau - k, \omega)$$

Using Equation 4.4 and remembering that

$$\chi(\tau + k, \omega) = 0$$

for $0 < \tau < k$ the above becomes

$$\psi(\tau, \omega) = e^{-j(n-1)\theta} \frac{\sin n\theta}{\sin \theta} \chi(\tau, \omega)$$

$$+ \frac{\chi(\tau - k, \omega)}{\sin \theta} \{e^{-j(n-2)\theta} \sin[(n-1)\theta] + e^{-j2(n-1)\theta} \sin \theta\}$$

Expanding the sine functions, the bracketed expression becomes

$$\frac{1}{2j} \{e^{-j(n-2)\theta} e^{j(n-1)\theta} - e^{-j(n-2)\theta} e^{-j(n-1)\theta}$$

$$+ e^{-j2(n-1)\theta} e^{j\theta} - e^{-j2(n-1)\theta} e^{-j\theta}\}$$

$$= \frac{e^{j\theta}}{2j} \{1 - e^{-j2n\theta}\} = e^{-j(n-1)\theta} \sin n\theta$$

Hence

$$\psi(\tau, \omega) = e^{-j(n-1)\theta} \frac{\sin n\theta}{\sin \theta} \{\chi(\tau, \omega) + \chi(\tau - k, \omega)\}$$

which is Equation 4.5.

4.3.3 Proof of Equation 4.6

In the range $-k < \tau < 0$

$$\psi(\tau, \omega) = \chi_1(\tau, \omega) + \chi(\tau + k, \omega)$$

Also $\chi(\tau - k, \omega) = 0$. Using Equation 4.4 the above becomes

$$\psi(\tau, \omega) = e^{-j(n-1)\theta} \frac{\sin n\theta}{\sin \theta} \chi(\tau, \omega)$$

$$+ \frac{\chi(\tau + k, \omega)}{\sin \theta} \{e^{-jn\theta} \sin[(n-1)\theta] + \sin \theta\}$$

The expanded bracketed expression becomes

$$\frac{1}{2j} \{e^{-jn\theta} e^{j(n-1)\theta} - e^{-jn\theta} e^{-j(n-1)\theta} + e^{j\theta} - e^{-j\theta}\}$$

$$= \frac{e^{j\theta}}{2j} \{1 - e^{-j2n\theta}\} = e^{-j(n-1)\theta} \sin n\theta$$

Hence

$$\psi(\tau, \omega) = e^{-j(n-1)\theta} \frac{\sin n\theta}{\sin \theta} \{\chi(\tau, \omega) + \chi(\tau + k, \omega)\}$$

which is Equation 4.6.

4.3.4 Proof of Equation 4.7

To derive Equations 4.7 and 4.8 it is necessary to set the repetition period $k = Ld$ and then use the formula

$$\chi(\tau, \omega) = \sum_{m=0}^{L-1} \chi_b(\tau - md, \omega) \mathscr{A}(m, \omega)$$

$$+ \sum_{m=1}^{L-1} e^{-jm\omega d} \chi_b(\tau + md, \omega) \mathscr{A}^*(m, -\omega)$$

in Equations 4.5 and 4.6. Over the range $0 < \tau < Ld$,

$$\chi_b(\tau + md, \omega) = 0$$

hence $\chi(\tau, \omega)$ becomes

$$\sum_{m=0}^{L-1} \chi_b(\tau - md, \omega) \mathscr{A}(m, \omega)$$

Similarly $\chi(\tau - k, \omega) = \chi(\tau - Ld, \omega)$ becomes

$$\chi_b(\tau - Ld, \omega) \mathscr{A}(0, \omega) + \sum_{m=1}^{L-1} e^{-jm\omega d} \chi_b(\tau + (m-L)d, \omega) \mathscr{A}^*(m, -\omega)$$

Changing the summation variable to $(L - m)$ allows $\chi(\tau - Ld, \omega)$ to be written

$$\chi_b(\tau - Ld, \omega) \mathscr{A}(0, \omega)$$

$$+ \sum_{m=1}^{L-1} e^{-j(L-m)\omega d} \chi_b(\tau - md, \omega) \mathscr{A}^*(L - m, -\omega)$$

Hence $\chi(\tau, \omega) + \chi(\tau - Ld, \omega)$ becomes

$$\{\chi_b(\tau, \omega) + \chi_b(\tau - Ld, \omega)\} \mathscr{A}(0, \omega)$$

$$+ \sum_{m=1}^{L-1} \chi_b(\tau - md, \omega)\{\mathscr{A}(m, \omega) + e^{-j(L-m)\omega d} \mathscr{A}^*(L - m, -\omega)\}$$

Since

$$\mathscr{A}(m, \omega) = \sum_{i=0}^{L-1-m} C_i C_{i+m}^* e^{-ji\omega d}$$

$$\mathscr{A}^*(L - m, -\omega) = \sum_{i=0}^{m-1} C_i^* C_{i+L-m} e^{-ji\omega d}$$

Changing the summation variable to $(i - L + m)$ gives

$$\mathscr{A}^*(L-m, -\omega) = \sum_{i=L-m}^{L-1} C_i C^*_{i-L+m} \, e^{-j(i-L+m)\omega d}$$

Since the sequence is periodic with length L, it follows that

$$C^*_{i-L+m} = C^*_{i+m}$$

Hence the expression for $\chi(\tau, \omega) + \chi(\tau - Ld, \omega)$ can be written

$$\{\chi_b(\tau, \omega) + \chi_b(\tau - Ld, \omega)\} \, \mathscr{A}(0, \omega) + \sum_{m=1}^{L-1} \chi_b(\tau - md, \omega)$$

$$\times \left\{ \sum_{i=0}^{L-1-m} C_i C^*_{i+m} \, e^{-ji\omega d} + \sum_{i=L-m}^{L-1} C_i C^*_{i+m} \, e^{-ji\omega d} \right\}$$

Substitution of the above in Equation 4.5 gives Equation 4.7.

4.3.5 Proof of Equation 4.8

Over the range $-Ld < \tau < 0$, $\chi_b(\tau - md, \omega) = 0$ hence $\chi(\tau, \omega)$ becomes

$$\chi_b(\tau, \omega) \, \mathscr{A}(0, \omega) + \sum_{m=1}^{L-1} e^{-jm\omega d} \chi_b(\tau + md, \omega) \, \mathscr{A}^*(m, -\omega)$$

Changing the summation variable to $(L - m)$ allows $\chi(\tau, \omega)$ to be written

$$\chi_b(\tau, \omega) \, \mathscr{A}(0, \omega) + \sum_{m=1}^{L-1} e^{-j(L-m)\omega d} \chi_b(\tau + (L-m)d, \omega) \, \mathscr{A}^*(L-m, -\omega)$$

$\chi(\tau + Ld, \omega)$ becomes

$$\sum_{m=0}^{L-1} \chi_b(\tau + (L-m)d, \omega) \, \mathscr{A}(m, \omega)$$

Hence $\chi(\tau, \omega) + \chi(\tau + Ld, \omega)$ becomes

$$\{\chi_b(\tau, \omega) + \chi_b(\tau + Ld, \omega)\} \, \mathscr{A}(0, \omega)$$

$$+ \sum_{m=1}^{L-1} \chi_b(\tau + (L-m)d, \omega) \{ \mathscr{A}(m, \omega) + e^{-j(L-m)\omega d} \mathscr{A}^*(L-m, -\omega)\}$$

which can be substituted in Equation 4.6 to give Equation 4.8.

4.4 THE MATCHED FILTER CONCEPT

Consider a filter with a real impulse response of $g(t)$ which is used to process a real signal $f(t)$. Assume that the filter output $h(t)$ has its peak value at $t = t_0$.

The convolution theorem (Reference 8, page 39) allows the peak output to be written in the form

$$h(t_0) = \int_{-\infty}^{\infty} f(x)g(t_0 - x)\,dx \qquad (4.9)$$

It will now be of interest to consider a second filter of impulse response $\mu f(t_m - t)$, where μ is a real constant. Assume that the output of this filter is $h_m(t)$ and has its peak value at $t = t_m$. Applying the convolution theorem once more, gives

$$h_m(t_m) = \int_{-\infty}^{\infty} f(x)\mu f(t_m - t_m + x)\,dx = \mu \int_{-\infty}^{\infty} f^2(x)\,dx \qquad (4.10)$$

The ratio of the peak outputs from the two filters is given by

$$\frac{h(t_0)}{h_m(t_m)} = \frac{\displaystyle\int_{-\infty}^{\infty} f(x)g(t_0 - x)\,dx}{\mu \displaystyle\int_{-\infty}^{\infty} f^2(x)\,dx}$$

If the Schwarz inequality [6] is applied to the integral in the numerator, the following relationship results

$$\left[\frac{h(t_0)}{h_m(t_m)}\right]^2 \leqslant \frac{\displaystyle\int_{-\infty}^{\infty} f^2(x)\,dx \int_{-\infty}^{\infty} g^2(t_0 - x)\,dx}{\mu^2 \left[\displaystyle\int_{-\infty}^{\infty} f^2(x)\,dx\right]^2} = \frac{\displaystyle\int_{-\infty}^{\infty} g^2(t_0 - x)\,dx}{\mu^2 \displaystyle\int_{-\infty}^{\infty} f^2(x)\,dx} \qquad (4.11)$$

Now

$$\int_{-\infty}^{\infty} g^2(t_0 - x)\,dx = \int_{-\infty}^{\infty} g^2(x)\,dx = \left(\begin{array}{l}\text{Energy of the impulse} \\ \text{response of } g(t)\end{array}\right)$$

If the gain factor μ is adjusted to make the energies of the two filter impulse responses equal, Equation 4.11 reduces to

$$|h(t_0)| \leqslant |h_m(t_m)|$$

Thus, if μ is selected to give equal filter impulse response energies the peak output of an arbitrary filter, with input $f(t)$, can never be greater than that from the filter having $\mu f(t_m - t)$ as its impulse response. Since

$$\mathscr{F}\{f(t + t_m)\} = e^{j\omega t_m} F(j\omega)$$

ıt follows that

$$\mathcal{F}\{f(t_m - t)\} = e^{-j\omega t_m} F(-j\omega)$$

The term $e^{-j\omega t_m}$ is the transfer function of a perfect delay line (its delay being t_m). As the inclusion of a delay line will not affect the magnitude of the peak value of the filter output, it is customary to refer to the filter having a transfer function $F(-j\omega)$ as the 'matched filter' for $f(t)$.

It follows that the impulse response of the matched filter is $f(-t)$ and (as $f(t)$ is assumed to be real) that its transfer function is given by either $F(-j\omega)$ or $F^*(j\omega)$.

It should be noted that it is always necessary to use delay in the realisation of matched filters for one-sided time functions. To illustrate this fact, let

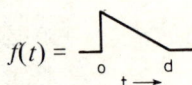

$$f(t) = $$

i.e. $f(t)$ is one-sided and time limited to d seconds. The impulse response of the corresponding matched filter will be given by $g(t) = f(-t)$, i.e.

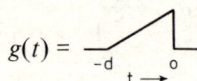

$$g(t) = $$

The above impulse response is unrealisable, since it is over before the impulse is applied! However a realisable impulse response with all of the properties of the matched filter would be $g(t - d)$, i.e.

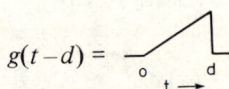

$$g(t-d) = $$

It can be seen that matched filters are only unrealisable in the case of non time-limited waveforms; such waveforms do not occur in practice, of course.

4.4.1 Noise properties of filters

If noise of constant spectral density (i.e. white noise) is applied to the input of a filter, the output noise power will be proportional to the area under the magnitude squared response of the filter. Thus

$$(\text{Output noise power}) \propto \int_{-\infty}^{\infty} |G(j\omega)|^2 \, d\omega$$

Parseval's theorem (Appendix 2) shows that the output noise power is also proportional to the energy of the filter impulse response. Thus

$$\text{(Output noise power)} \propto \int_{-\infty}^{\infty} g^2(t)\, dt$$

The above facts lead to the concept of the noise bandwidth of a filter. The noise bandwidth of a low-pass filter is defined as the bandwidth of a rectangular filter having the same d.c. gain, and a magnitude squared response which encloses the same area as the magnitude squared response of the low pass filter.

Since the area under the magnitude squared response of a unity gain rectangular filter of bandwidth B is given by $2B$, the formula for the noise bandwidth of a filter of transfer function $G(j\omega)$ is

$$\left\{\begin{matrix} \text{Noise bandwidth} \\ \text{(rad/s)} \end{matrix}\right\} = \frac{\int_{-\infty}^{\infty} |G(j\omega)|^2\, d\omega}{2|G(0)|^2} = \frac{\pi \int_{-\infty}^{\infty} g^2(t)\, dt}{|G(0)|^2} \qquad (4.12)$$

The peak signal-to-noise ratio at the output of a filter having an impulse response $g(t)$ can be defined as

$$\left(\frac{S}{N}\right) = \frac{h^2(t_0)}{\int_{-\infty}^{\infty} g^2(t)\, dt} \qquad (4.13)$$

where $h(t_0)$ is the peak value of the output due to $f(t)$ at the input. This definition makes (S/N) proportional to the ratio of peak signal power to noise power.

Now consider a matched filter with a peak signal-to-noise ratio of $(S/N)_m$. Assume that the filter impulse response is given by $\mu f(-t)$ and that μ has been adjusted to equalise the impulse response energies, i.e. so that

$$\mu^2 \int_{-\infty}^{\infty} f^2(-t)\, dt = \int_{-\infty}^{\infty} g^2(t)\, dt$$

If the resulting peak output signal is $h_m(t_m)$, the output peak signal-to-noise ratio will be given by

$$\left(\frac{S}{N}\right)_m = \frac{h_m^2(t_m)}{\mu^2 \int_{-\infty}^{\infty} f^2(-t)\, dt} = \frac{h_m^2(t_m)}{\int_{-\infty}^{\infty} g^2(t)\, dt}$$

Combining this result with Equation 4.13 gives

$$\left(\frac{S}{N}\right) = \frac{h^2(t_0)}{h_m^2(t_m)} \left(\frac{S}{N}\right)_m \qquad (4.14)$$

As it has already been shown that $h^2(t_0)$ can never be greater than $h_m^2(t_m)$, it follows that

$$\left(\frac{S}{N}\right) \leqslant \left(\frac{S}{N}\right)_m$$

Thus if a signal, embedded in white noise, is applied to an arbitrary filter the output peak signal-to-noise ratio will never be greater than the peak signal-to-noise ratio at the output of the matched filter.

4.5 THE MATHEMATICAL TREATMENT OF THE MATCHED FILTER RECEIVER

This section is used to investigate the RF envelope of the output of the receiver discussed in Section 4.1 (Fig. 4.1). The transmitted signal is assumed to be

$$f(t, \omega_0) = |a(t)| \cos[\omega_0 t + \phi(t)]$$

and the return signal a delayed and Doppler shifted version of the transmitted signal, i.e.

$$f[(t-x), (\omega_0 + y)]$$

The main mixer of the receiver is assumed to have a single (lower) sideband output of

$$f[(t-x), (\omega_0 + y - \omega_v)]$$

It will also be assumed that $(\omega_0 - \omega_v)$ is high enough for the exponential approximation to the complex analytic signal to apply (see Section 7.4).

It will be shown that the output RF envelope of an arbitrary filter, having a complex analytic form of $B[j(\omega - \omega_1)]$ is given by

$$|h_a(t)| = \tfrac{1}{2} \left| \int_{-\infty}^{\infty} a(z)b(t-x-z) \, e^{j(\omega_0 - \omega_1 - \omega_v + y)z} \, dz \right| \qquad (4.15)$$

It is also shown that the impulse and frequency responses of the complex analytic filter corresponding to the matched filter for $f(t)$ are given by

$$f_a^*(-t) \quad \text{and} \quad F_a^*(j\omega)$$

For the matched filter case the receiver output RF envelope is given by

$$|h_a(t)| = \tfrac{1}{2} |\psi[(t-x), (y + \omega_0 - \omega_v - \omega_1)]| \qquad (4.16)$$

where $|\chi(\tau, \omega)|$ is the uncertainty function derived from $f(t)$.

4.5.1 Derivation of results

From the above discussion it will be seen that the impulse response of the arbitrary complex analytic filter is given by $b(t) e^{j\omega_1 t}$. Also the complex analytic signal corresponding to the filter input is

$$a(t-x) e^{j(\omega_0 + y - \omega_v)(t-x)}$$

where

$$a(t) = |a(t)| e^{j\phi(t)}$$

Hence, using Section 7.6 (Fig. 7.7d) and the convolution theorem, the complex analytic form of the receiver output is given by

$$h_a(t) = \frac{1}{2} \int_{-\infty}^{\infty} a(z-x) e^{j(\omega_0 + y - \omega_v)(z-x)} b(t-z) e^{j(t-z)\omega_1} dz$$

which after a change of dummy variable reduces to

$$h_a(t) = \frac{1}{2} e^{j(t-x)\omega_1} \int_{-\infty}^{\infty} a(z) b(t-x-z) e^{j(\omega_0 - \omega_1 - \omega_v + y)z} dz \quad (4.17)$$

The RF envelope is given by $|h_a(t)|$ (see Section 7.3), hence Equation 4.15 follows.

If the arbitrary filter is now replaced by a matched filter of impulse response $g(t)$ it follows from Section 4.4 that

$$g(t) = f(-t) \quad (4.18)$$

$$G(j\omega) = F(-j\omega) = F^*(j\omega) \quad (4.19)$$

To find the complex analytic filter corresponding to $G(j\omega)$ note that

$$f(t) = \text{Re}\{f_a(t)\} = \frac{1}{2} [f_a(t) + f_a^*(t)]$$

Hence, using Section 6.7

$$F(j\omega) = \frac{1}{2}\{F_a(j\omega) + F_a^*(-j\omega)\}$$

which with Equation 4.19 gives

$$G(j\omega) = \frac{1}{2}\{F_a^*(j\omega) + F_a(-j\omega)\}$$

Since, by definition, $F_a(j\omega) = 0$, for $\omega < 0$, and $F_a(-j\omega) = 0$, for $\omega > 0$, Section 7.6 (Equation 7.42) leads to

$$G_a(j\omega) = F_a^*(j\omega) \quad (4.20)$$

Note particularly that $F_a^*(j\omega)$ is *not* equal to $F_a(-j\omega)$, since $f_a(t)$ is complex.

Using Section 6.7 (Equation 6.17) with Equation 4.20 gives

$$g_a(t) = f_a^*(-t) \qquad (4.21)$$

It follows from Equation 4.21 that the output from a matched filter receiver can be obtained from Equation 4.17 by replacing $b(z)$ by $a^*(-z)$, hence

$$h_a(t) = \tfrac{1}{2} e^{j(t-x)\omega_1} \int_{-\infty}^{\infty} a(z)a^*(z+x-t)\, e^{j(\omega_0-\omega_1-\omega_v+y)z}\, dz \qquad (4.22)$$

The integral in Equation 4.22 will be recognised (Equation 2.14) as $\chi\left[-(t-x),\ -(\omega_0 - \omega_1 - \omega_v + y)\right]$. Hence, the output RF envelope is given by

$$|h_a(t)| = \tfrac{1}{2}|\chi\left[(t-x),\ (\omega_0 - \omega_1 - \omega_v + y)\right]| \qquad (4.23)$$

Equation 4.23 follows from 4.22 by virtue of Section 2.2 (Equation 2.18).

4.6 PRACTICAL FOURIER TRANSFORM CALCULATORS

The calculator used in the Fourier transform receiver is required to find

$$\int_0^{nk} f(t)\, e^{-j\omega t}\, dt$$

at a selected number of values of ω. The parameters n and k represent the number of processed pulses and the pulse repetition frequency period, respectively. Once one batch of pulses has been processed it is required that the integrators be reset, ready to process the next batch.

The direct method of implementing the calculator would be to expand the above integral to give

$$\int_0^{nk} f(t)\cos(\omega t)\, dt - j \int_0^{nk} f(t)\sin(\omega t)\, dt$$

Since it is the modulus (or the square of the modulus) of the above sum which is required, a direct implementation method would be by the system illustrated in Fig. 4.17.

If precision requirements dictated calculation at m frequencies it would either be necessary to use m parallel channels (each with a different value of ω), for a fast read out, or to take m times as long by using one channel m times.

The system of Fig. 4.17 could be replaced by any of the well known systems used for spectrum analysis. Owing to the rapid advances in integrated circuit techniques a particularly attractive method might be to sample the output signal from the receiver multiplier and then feed

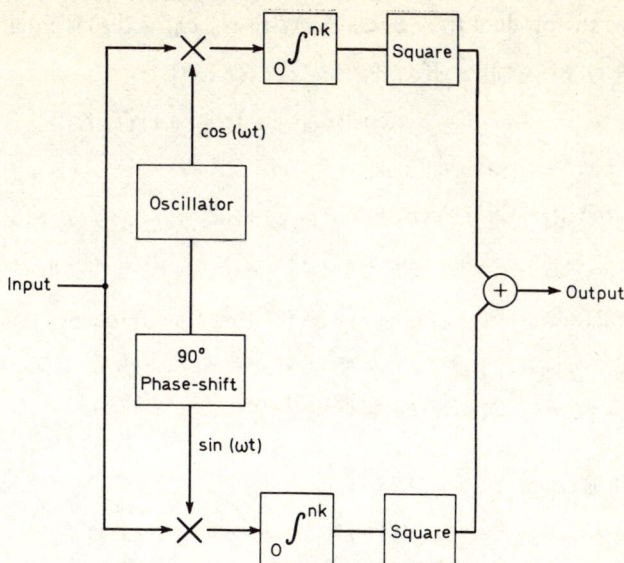

Fig. 4.17. Direct implementation of F.T. calculator

the samples to a digital system for calculation. A parallel method of calculation, making the most of finite logic speeds, would be one of the FFT algorithms discussed in Section 6.14.

The classic spectrum analysis scheme, using a narrow band IF amplifier in conjunction with a swept local oscillator is essentially a series scheme since the oscillator must sweep slowly for accurate results. The equivalent parallel scheme would employ a bank of narrow band filters each tuned to a different frequency. The filter bandwidths should be approximately equal to $1/(nk)$ Hz.

4.7 THE MATHEMATICAL TREATMENT OF THE FOURIER TRANSFORM RECEIVER

With reference to Fig. 4.7 the transmitted signal is defined as

$$f(t, \omega_0) = |a(t)| \cos[\omega_0 t + \phi(t)] = \text{Re}\{a(t)\, e^{j\omega_0 t}\}$$

The inputs to the main mixer are

$$f(t-x, \omega_0 + y) \quad \text{and} \quad f(t-t_v, \omega_0 - \omega_L)$$

Hence the product $f(t-x, \omega_0 +y) f(t-t_v, \omega_0 -\omega_L)$ is equal to

$$|a(t-x)|\,|a(t-t_v)|\,\cos[(\omega_0+y)(t-x)+\phi(t-x)]$$
$$\cos[(\omega_0-\omega_L)(t-t_v)+\phi(t-t_v)]$$

$$= \tfrac{1}{2}|a(t-x)|\,|a(t-t_v)|$$

$$\{\cos[(y+\omega_L)t + \phi(t-x) - \phi(t-t_v) + (\omega_0-\omega_L)t_v -(\omega_0+y)x]$$
$$+ \cos[(2\omega_0-\omega_L+y)t + \phi(t-x) + \phi(t-t_v) -(\omega_0-\omega_L)t_v-(\omega_0+y)x]\}$$

The mixer output is assumed to be the first (low frequency) term i.e.

$$\tfrac{1}{4}|a(t-x)|\,|a(t-t_v)|\,e^{j\phi(t-x)}\,e^{-j\phi(t-t_v)}\,e^{j(y+\omega_L)t}\,e^{j(\omega_0-\omega_L)t_v}\,e^{-j(\omega_0+y)x}$$
$$+ \tfrac{1}{4}|a(t-x)|\,|a(t-t_v)|\,e^{-j\phi(t-x)}\,e^{j\phi(t-t_v)}\,e^{-j(y+\omega_L)t}$$
$$e^{-j(\omega_0-\omega_L)t_v}\,e^{j(\omega_0+y)x}$$

which is equal to

$$\tfrac{1}{4}\{a(t-x)a^*(t-t_v)\,e^{j(y+\omega_L)t}\,e^{j(\omega_0-\omega_L)t_v}\,e^{-j(\omega_0+y)x}$$
$$+ a^*(t-x)a(t-t_v)\,e^{-j(y+\omega_L)t}\,e^{-j(\omega_0-\omega_L)t_v}\,e^{j(\omega_0+y)x}\} \qquad (4.24)$$

But

$$\mathscr{F}\{a(t-x)a^*(t-t_v)\} = \int_{-\infty}^{\infty} a(t-x)a^*(t-t_v)\,e^{-j\omega t}\,dt$$

$$= \int_{-\infty}^{\infty} a(t)a^*(t+x-t_v)\,e^{-j(t+x)\omega}\,dt \qquad (4.25)$$

$$= e^{-j\omega x}\,\chi\,[x-t_v,\,\omega]$$

Similarly

$$\mathscr{F}\{a^*(t-x)a(t-t_v)\} = e^{-j\omega t_v}\,\chi\,[t_v-x,\,\omega] \qquad (4.26)$$

where

$$\chi(\tau,\,\omega) = \int_{-\infty}^{\infty} a(t)a^*(t+\tau)\,e^{-j\omega t}\,dt$$

Using Equations 4.25 and 4.26 the Fourier transform of 4.24 can be written

$$\tfrac{1}{4}\{e^{j(\omega_0-\omega_L)t_v}\,e^{-j(\omega+\omega_0-\omega_L)x}\,\chi\,[x-t_v,\,\omega-y-\omega_L]$$
$$+ e^{-j(\omega+\omega_0+y)t_v}\,e^{j(\omega_0+y)x}\,\chi\,[t_v-x,\,\omega+y+\omega_L]\} \qquad (4.27)$$

For a real bandpass function $h(t)$

$$\mathscr{F}\{h(t)\} = H(j\omega) = |H(j\omega)|\,e^{j\theta(j\omega)}$$

Hence

$$h(t) = \frac{1}{2\pi} \int_{-\infty}^{\infty} H(j\omega) e^{j\omega t} d\omega = \int_{0}^{\infty} 2|H(j\omega)| \cos[\omega t + \theta(j\omega)] df$$

If $h(t)$ were applied to a physical 'Fourier transform calculator' (e.g. a spectrum analyser) the output waveform would represent the amplitude distribution of the various cosine components, measured in V/Hz, i.e. $2|H(j\omega)|$.

It can be seen, from the above remarks, that the output from the Fourier transform receiver will be given by twice the modulus of the expression 4.27, when evaluated for $\omega > 0$.

If ω_L is sufficiently high, the contribution from $\chi[t_v - x, \omega + y + \omega_L]$ will be negligible leading to an output waveform given by

$$\tfrac{1}{2}|\chi[x - t_v, \omega - y - \omega_L]|$$

4.7.1 The effect of weighting

The effect of time weighting in a Fourier transform receiver will now be compared with that of frequency weighting in a matched filter receiver.

To investigate the effect of multiplying the signal at the input of the Fourier transform calculator (Fig. 4.7) by the function $m(t - t_v)$, it is necessary to put $m(t - t_v)$ as a multiplier before expression 4.24. The resultant expression for the Fourier transform modulus can be written as

$$\tfrac{1}{2}\left| \int_{-\infty}^{\infty} a(t_v - x - z)a^*(-z)m(-z) e^{-j(\omega_L + y - \omega)z} dz \right| \qquad (4.28)$$

The effect of frequency weighting in a matched filter receiver can be investigated by assuming that the complex analytic impulse response corresponding to the receiver filter is $a^*(-t)c(t) e^{j\omega_1 t}$ rather than $a^*(-t) e^{j\omega_1 t}$. Replacing $b(t - x - z)$ in Section 4.5 (Equation 4.15) by $a^*(z + x - t)c(t - x - z)$ and subsequent simplification gives the output RF envelope as

$$\tfrac{1}{2}\left| \int_{-\infty}^{\infty} a(t - x - z)a^*(-z)c(z) e^{-j(\omega_0 - \omega_1 + y - \omega_v)z} dz \right| \qquad (4.29)$$

Expressions 4.28 and 4.29 are discussed in Section 4.2.1.

Chapter 5

SOME EXAMPLES OF SIGNAL PROCESSING METHODS

Examples are given in this chapter of the methods used to process two types of modulated signal. The resulting precision and resolution are discussed in terms of the quantitative definitions given in Section 1.4.

5.1 THE RECTANGULAR PULSE WITH CONSTANT CARRIER FREQUENCY

A rectangular pulse with constant carrier frequency can be represented by

$$f(t) = \left[\underset{\substack{\text{o} \quad \text{d} \\ t \longrightarrow}}{\rule{0pt}{0pt}\sqcap\sqcap} \right] \cos(\omega_0 t) \qquad (5.1)$$

Hence, for ω_0 sufficiently high

$$f_d(t) = \left[\underset{\substack{\text{o} \quad \text{d} \\ t \longrightarrow}}{\rule{0pt}{0pt}\sqcap\sqcap} \right] e^{j\omega_0 t} \qquad (5.2)$$

The appropriate uncertainty function was evaluated in Section 3.1 where it was shown (Equations 3.10 and 3.11) that the inherent range and velocity precisions are

$$r = 0.5\, cd \qquad (5.3)$$

$$v = 0.6\, \frac{c}{f_0 d} \qquad (5.4)$$

This performance can be achieved by using an unweighted matched filter or Fourier transform receiver.

5.1.1 The Fourier transform receiver

The Fourier transform receiver of Fig. 4.7 is often used with this waveform in practical radar systems. Sometimes the delay t_v is set at a constant value and the system used to measure the velocities of targets in a fixed range bracket. The Fourier transform calculator is often implemented as a bank of narrowband filters each tuned to a different centre frequency.

To design such a system, given a specification of the desired range and velocity precision, one should choose the pulse length from Equation 5.3 and the carrier frequency from Equation 5.4. This procedure will allow the fastest readout of velocity information—the required Fourier transform integration time being equal to the pulse duration d. In the case of a filter bank the filter bandwidth (in Hz) should be equal to about $1/d$.

Equation 3.5 gives

$$|\chi(0, \omega)| = d \left| \frac{\sin(\omega d/2)}{(\omega d/2)} \right| \tag{5.5}$$

To avoid loss of precision due to the interference of the two uncertainty functions corresponding to a given target (see Fig. 4.8) it is necessary to choose an offset frequency ω_L such that $|\chi(0, \omega)|$ is very small for ω equal to $2\omega_L$.

The expression in Equation 5.5 will be more than 60 dB down for $\omega d > 2000$, thus a suitable choice for ω_L (and hence the IF frequency) would be

$$\omega_L \geqslant \frac{1000}{d} \tag{5.6}\dagger$$

The pulse repetition frequency period k would be chosen high enough to eliminate second time round errors.

Although it is theoretically possible to carry out the above procedure, in many cases the required carrier frequency will be too high for practical implementation. Combining Equations 5.3 and 5.4 gives the required carrier frequency as

$$f_0 = \frac{0.3\, c^2}{vr} \tag{5.7}$$

Taking an upper practical limit of 100 GHz for f_0 (in terms of lasers the upper limit could be as high as 10^6 GHz), gives

$$vr \geqslant 0.1 \tag{5.8}$$

The units of v and r are mile/s and miles respectively.

By way of example, a precision specification of 1 mile and 360 mph would lead to

$$d = 10.8\ \mu s, \quad f_0 = 100\ \text{GHz}, \quad f_L \geqslant 15\ \text{MHz}$$

---

† It is usual to employ a lower value of IF (ω_L) than that indicated by Equation 5.6. This is permissible because the $(\sin x)/x$ expression of Equation 5.5 applies to a pulse with zero rise and fall times.

Each filter in the filter bank would require a bandwidth of about 100 kHz.

If the required carrier frequency cannot be achieved a combined range and velocity precision specification can be satisfied by using the improvement in Doppler precision due to repetition. To exploit the precision improvement it is only necessary to increase the calculator integration time to nk, which gives an integration over n pulses. The filter bank bandwidths should be reduced to approximately $1/(nk)$ Hz. The new expression for velocity precision (replacing Equation 5.4) is

$$v = 0.6 \frac{c}{f_0 nk}, \qquad n \gg 1 \tag{5.9}$$

Although Equation 5.9 indicates that velocity precision can be improved indefinitely by lengthening the integration time it should be noted that the readout time becomes progressively longer, hence changes in velocity will take longer to become apparent.

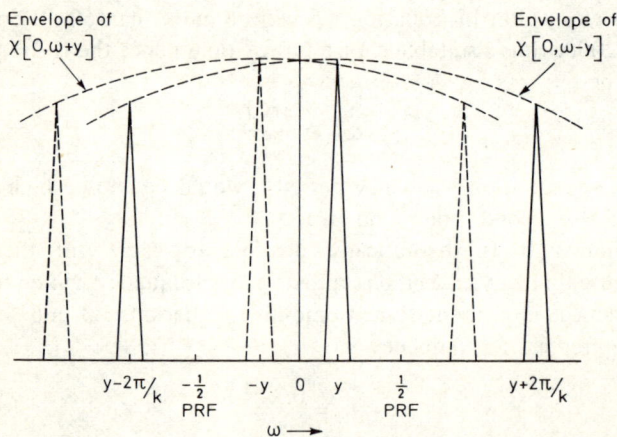

Fig. 5.1. The sum of $\chi(0, \omega - y)$ and $\chi(0, \omega + y)$ when both are multiplied by the weighting function due to repetition

If one is relying upon the Doppler 'bar' effect to give velocity precision the system can be simplified by omitting the offset oscillator—leading to a homodyne receiver—the price paid is a loss of knowledge of the sign of velocity changes. Reference to Fig. 5.1 will show that, for this special case, omission of the offset oscillator does not lead to a loss of precision as long as the maximum Doppler frequency is less than half the pulse repetition frequency.

5.1.2 The matched filter receiver

It will be seen from the above discussion that the implementation of a Fourier transform receiver for a constant carrier rectangular pulse is relatively straightforward. The physically realisable form of the matched filter receiver would need a filter with an impulse response given by a suitably delayed version of $f(-t)$. Reference to Equation 5.1 shows that, for this particular waveform, such an impulse response would be given by the expression for $f(t)$ itself. The corresponding filter transfer function would be given by

$$\tfrac{1}{2}\{G^*[j(-\omega-\omega_0)]+G[j(\omega-\omega_0)]\} \tag{5.10a}$$

where

$$G(j\omega)=e^{-j(\omega d/2)}\,\mu d\left[\frac{\sin(\omega d/2)}{\omega d/2}\right] \tag{5.10b}$$

The matched filter would actually be implemented at a suitably high IF frequency ω_I–rather than at ω_0–and used in the system of Fig. 4.1. It turns out that it is not particularly easy to construct such a filter (a method is given by Skolnik [14]) so most practical systems would use a non-matched filter in the Fig. 4.1 scheme.

As an example of a non-matched system the performance using a single tuned circuit type of filter will now be considered. It is assumed that the complex analytic form of the filter transfer function is

$$B[j(\omega-\omega_1)] \tag{5.11a}$$

where

$$B(j\omega)=\frac{a}{a+j\omega} \tag{5.11b}$$

The corresponding baseband impulse response is

$$b(t)=a\,e^{-at}\,u(t) \tag{5.12}$$

It is shown below that use of the above filter leads to a receiver output envelope of the same shape as a cut (parallel to the τ axis) through the following 3-dimensional shape

$$|\psi(\tau,\omega)|=\begin{cases} 0, & \tau<0 \\[2ex] \dfrac{ad}{2\sqrt{(ad)^2+(\omega d)^2}}\left[1+e^{-2ad(\tau/d)}-2\,e^{-ad(\tau/d)}\cos\left\{\omega d\left(\dfrac{\tau}{d}\right)\right\}\right]^{\frac{1}{2}}, \\ & 0<\tau<d \\[2ex] \dfrac{ad\,e^{-ad(\tau/d)}}{2\sqrt{(ad)^2+(\omega d)^2}}\left[1+e^{2ad}-2\,e^{ad}\cos(\omega d)\right]^{\frac{1}{2}}, & \tau>d \end{cases}$$

To evaluate the above expression it is necessary to decide upon a value for the product *ad*. It is common practice to make the 3 dB bandwidth of the filter (in Hz) equal to the reciprocal of the pulse width. The 3 dB bandwidth of the RF filter is equal to 2*a* rad/s, thus the above philosophy implies *ad* = π.

An alternative approach would be to use a filter having the same noise bandwidth as the matched filter. It is shown below that this implies *ad* = 2.

Figures 5.2 and 5.3 show the effects of the above choices of filter bandwidth upon the delay and Doppler performance; matched filter responses are also shown for comparison. The results are summarised in Table 5.1 in which the delay figures are values of τ/d and the Doppler figures are values of $\omega d/2\pi$.

It can be seen that either choice of bandwidth leads to results which are not too different from those obtained with a matched filter. Neither

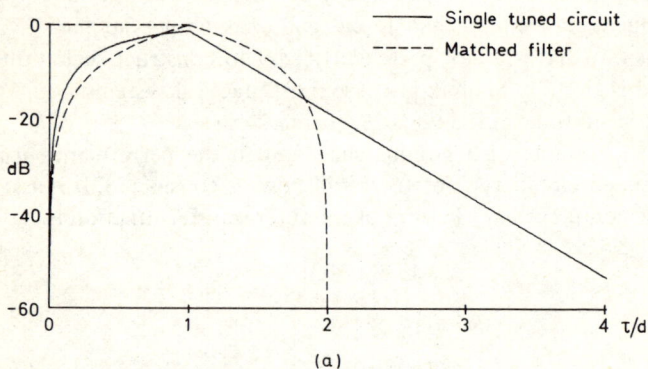

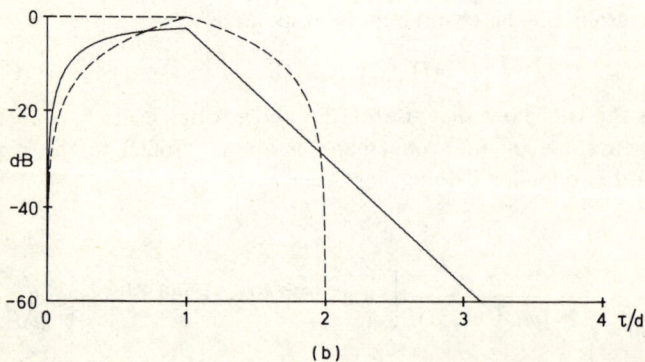

Fig. 5.2. Delay   performance   (a) ad = 2   *(i.e.   IF   bandwidth = 2/*π*d Hz) (b)* ad = π *(i.e. IF bandwidth = 1/d Hz)*

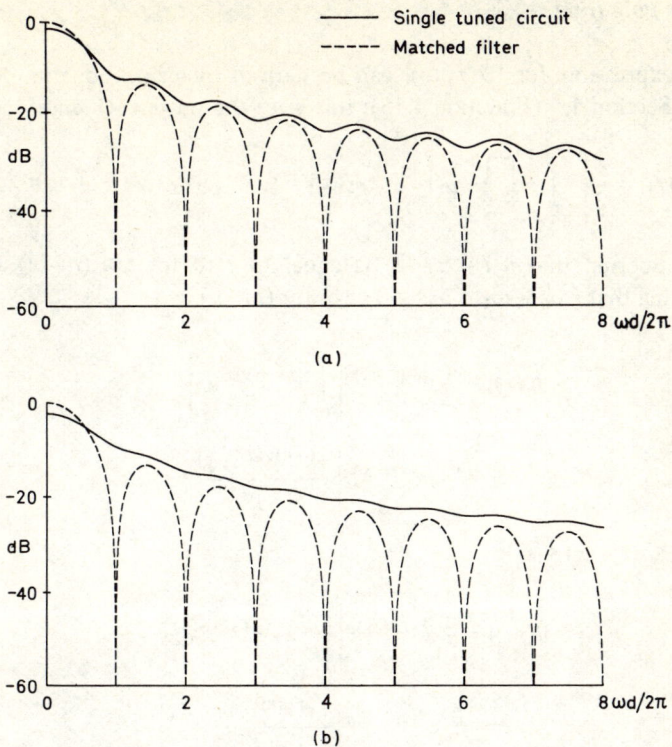

Fig. 5.3. Doppler performance (a) ad = 2 (i.e. IF bandwidth = 2/πd Hz) (b) ad = π (i.e. IF bandwidth = 1/d Hz)

Table 5.1

	Matched filter	*ad* = 2	*ad* = π
Delay precision	1	1.05	1
60 dB delay resolution	2	4.5	3.2
Doppler precision	1.2	1.4	1.6
Comparative *S/N* (dB)	0	−1.26	−2.34

the matched filter nor the single tuned circuit will give good Doppler resolution, since both $|\chi(\tau, \omega)|$ and $|\psi(\tau, \omega)|$ fall off as $1/\omega$ for large values of ω.

5.1.3 Derivation of results

The equations leading to Figs 5.2 and 5.3 will now be derived.

Single pole filter

The expression for $|\psi(\tau, \omega)|$ can be derived by using Equation 5.12 with Section 4.5 (Equation 4.15); this gives the output RF envelope as

$$|h_d(t)| = \frac{a}{2}\left| \int_{-\infty}^{\infty} \underset{\underset{z \to}{0 \quad d}}{\sqcap} u(t-x-z)\, e^{-(t-x-z)a}\, e^{j(\omega_0 - \omega_1 - \omega_v + y)z}\, dz \right|$$

Remembering that $u(t - x - z)$ is equal to zero for $z > (t - x)$ and assuming that $(\omega_0 - \omega_1 - \omega_v) = 0$ gives for $(t - x) > d$

$$|h_d(t)| = \frac{a}{2}\, e^{-(t-x)a}\left| \int_{0}^{d} e^{(a+jy)z}\, dz \right|$$

$$= \frac{a}{2}\, e^{-(t-x)a}\left| \frac{e^{(a+jy)d} - 1}{a + jy} \right|$$

For $0 < (t - x) < d$

$$|h_d(t)| = \frac{a}{2}\, e^{-(t-x)a}\left| \int_{0}^{t-x} e^{(a+jy)z}\, dz \right|$$

$$= \frac{a}{2}\, e^{-(t-x)a}\left| \frac{e^{(a+jy)(t-x)} - 1}{a + jy} \right|$$

For $(t - x) < 0$

$$h_d(t) = 0$$

Putting $(t - x) = \tau$ and $y = \omega$ and simplifying the above expressions leads to the expression for $|\psi(\tau, \omega)|$ quoted in Section 5.1.2.

Since

$$\int_{-\infty}^{\infty} b^2(t)\, dt = a^2 \int_{0}^{\infty} e^{-2at}\, dt = \frac{a}{2} \tag{5.13}$$

the filter noise bandwidth will be given by

$$\left\{ \begin{array}{l} \text{Noise} \\ \text{Bandwidth} \end{array} \right\} = \frac{\pi \int_{-\infty}^{\infty} b^2(t)\, dt}{|B(o)|^2} = \frac{\pi a}{2} \tag{5.14}$$

Equation 5.14 follows from Section 4.4, Equation 4.12.

Matched filter

The baseband impulse response of the matched filter is given by

$$b(t) = \underbrace{\qquad}_{\substack{0 \qquad d \\ t \longrightarrow}} \overline{} \mu$$

Hence

$$\int_{-\infty}^{\infty} b^2(t)\, dt = \mu^2 d$$

To equalise the impulse response energies of the two filters it is necessary to put $\mu^2 d$ equal to $a/2$, this gives

$$\mu = \sqrt{\left(\frac{a}{2d}\right)} \qquad (5.15)$$

The matched filter d.c. gain is given by

$$B(o) = \int_{-\infty}^{\infty} b(t)\, dt = \mu d = \sqrt{\left(\frac{ad}{2}\right)}$$

The expression for noise bandwidth is

$$\left\{\begin{matrix} \text{Noise} \\ \text{Bandwidth} \end{matrix}\right\} = \frac{\pi \int_{-\infty}^{\infty} b^2(t)\, dt}{|B(o)|^2} = \frac{\pi}{d} \text{ rad/s} \qquad (5.16)$$

The noise bandwidths of the two filters can be equalised by combining Equations 5.14 and 5.16 to give $ad = 2$.

The output envelope of the matched filter receiver can be found either by substituting the filter baseband impulse response in Equation 4.15 or by evaluating $\mu/2|\chi(\tau - d, \omega)|$ from Equation 3.3. The result is

$$\left\{\begin{matrix} \text{Matched filter} \\ \text{output envelope} \end{matrix}\right\} = \begin{cases} \left|\sqrt{\left(\frac{a}{2\omega^2 d}\right)}\right| \sin\left[\frac{\omega d}{2}\left(1 - \left|\left(\frac{\tau}{d}\right) - 1\right|\right)\right], & 0 < \tau < 2d \\ 0, & \text{otherwise.} \end{cases}$$

As the filter gains have been adjusted to give equal impulse response energies, the peak signal-to-noise ratios can be compared by comparing the peak output signals (see Section 4.4.1).

Both filters will give their peak outputs for ω equal to zero and τ equal to d. The peak output from the matched filter will be $\sqrt{(ad/8)}$, while the peak output from the single pole filter will be

$$\tfrac{1}{2} e^{-ad} \sqrt{(1 + e^{2ad} - 2 e^{ad})}$$

Using these results with Equation 4.14 gives

$$\frac{\left(\dfrac{S}{N}\right) \text{ for single pole filter}}{\left(\dfrac{S}{N}\right) \text{ for matched filter}} = \frac{2}{ad}\left[1 + e^{-2ad} - 2\,e^{-ad}\right]$$

5.2 THE RECTANGULAR PULSE WITH SAWTOOTH FM

A rectangular pulse with sawtooth† FM can be represented by

$$f(t) = \left[\underset{\text{o} \quad \text{d}}{\rule{0pt}{0pt}}\rule{0pt}{0pt}\right]\cos\left[\omega_0 t + \frac{\Delta}{2d}t^2\right] \tag{5.17}$$

Δ is the total change in frequency (rad/s) during the pulse. The 'dispersion factor' of such a pulse is often mentioned, this is defined as the dimensionless product $d\Delta$.

If ω_0 is sufficiently high

$$f_a(t) = \left[\underset{\text{o} \quad \text{d}}{\rule{0pt}{0pt}}\rule{0pt}{0pt}\right] e^{j(t^2\Delta/2d)}\, e^{j\omega_0 t} \tag{5.18}$$

The uncertainty function of this waveform is discussed and illustrated in Section 3.2. The inherent precision relationships (for $d\Delta$ high) are

$$\frac{\omega d}{2\pi} = 1.2 \tag{5.19}$$

$$\frac{\tau\Delta}{2\pi} = 1.2 \tag{5.20}$$

Alternatively

$$v = \frac{0.6c}{f_0 d} \tag{5.21}$$

$$r = \frac{0.6c}{\Delta/2\pi} \tag{5.22}$$

5.2.1 The Fourier transform receiver

The Fourier transform receiver of Fig. 4.7 will exhibit the above characteristics together with the poor resolution predicted by the uncertainty function. Resolution can be improved by time weighting the output of the main mixer before the Fourier transform is

† i.e. successive pulses have the same initial frequency.

calculated. This effect will now be discussed, together with a fuller treatment of the effects of delay–Doppler coupling which were discussed qualitatively in Section 4.2.

It was shown in Section 4.7 that, provided the offset frequency ω_L was high enough, the output of the Fourier transform receiver was a cut through the modulus of a displaced uncertainty function, specifically

$$h(\omega) = \tfrac{1}{2}|\chi[(x-t_v),(\omega-y-\omega_L)]| \tag{5.23}$$

where $h(\omega)$ is the receiver output function and the other variables are defined in Fig. 4.7. In practice the receiver output will be a function of time but the variable is left as ω to avoid confusion later on.

Expanding Equation 5.23 by means of Equation 2.15 gives

$$h(\omega) = \tfrac{1}{2}\left| \mathscr{F}\{e(t)\, e^{j(t^2\Delta/2d)}\, e(t+x-t_v)\, e^{-j(\Delta/2d)(t+x-t_v)^2}\}\right|_{\omega \to \omega - y - \omega_L}$$

where

$$e(t) = \underset{\underset{t \longrightarrow}{0 \qquad d}}{\rule{0pt}{0pt}\sqcup\!\!\sqcap^{--1}}$$

Multiplying out the exponential terms gives

$$h(\omega) = \tfrac{1}{2}\left| \mathscr{F}\{e(t)\, e(t+x-t_v)\, e^{-j(t\Delta/d)(x-t_v)}\}\right|_{\omega \to (\omega - y - \omega_L)}$$

Hence

$$h(\omega) = \left| C\left[j\left(\omega + \frac{\Delta}{d}(x-t_v) - y - \omega_L\right)\right]\right| \tag{5.24}$$

where

$$C(j\omega) = \tfrac{1}{2}\,\mathscr{F}\{e(t)\, e(t+x-t_v)\} \tag{5.25}$$

As was shown in Section 4.7, the complete receiver output waveform is given by the above expression together with another symmetrically displaced uncertainty function. This provides the 'balancing' term necessary for the Fourier transform of a real function. It is usual to set $(x - t_v)$ equal to zero and make the offset frequency ω_L high enough to ensure separation of the two uncertainty functions. It can be seen that, for this special case, one of the effects of target delay x is to cause a shifting of $C(j\omega)$ along the ω axis proportional to the frequency sweep Δ. If the variable delay and offset oscillator were dispensed with Equation 5.24 would become

$$h(\omega) = \left| C\left[j\left(\omega + \frac{x\Delta}{d} - y\right)\right]\right| \tag{5.26}$$

As long as $x\Delta$ is large enough to ensure that $h(\omega)$ has negligible amplitude at low frequencies this function will have enough inherent offset to overcome interference from its 'mate'.

It can be seen from Equation 5.26 that changes in target delay x and Doppler offset y have the same shifting effect upon the output waveform. The delay-Doppler coupling is not complete since changing delay has an effect on the shape of $C(j\omega)$ (Equation 5.25). It can easily be shown, however, that for large $d\Delta$ products the change in shape will be negligible. Since

$$e(t) = \begin{array}{c} \underset{\text{o} \quad \text{d}}{\rule{0pt}{1em}} \\ t \longrightarrow \end{array} \rule[0.5em]{0pt}{0pt}^{-1}$$

it follows that

$$e(t+x) = \begin{array}{c} \underset{-x \quad d-x}{\rule{0pt}{1em}} \\ t \longrightarrow \end{array}{}^{-1}$$

Hence

$$e(t)\,e(t+x) = \begin{array}{c} \underset{\text{o} \quad d-x}{\rule{0pt}{1em}} \\ t \longrightarrow \end{array}{}^{-1} \tag{5.27}$$

Thus the length of the $e(t)\,e(t+x)$ pulse becomes

$$d\left[1 - \frac{x\Delta}{d\Delta}\right] \tag{5.28}$$

The question of change of pulse shape with change of target delay can now be resolved by comparing the Fourier transform of a pulse of length d with a pulse whose length is given by Equation 5.28. Clearly, for moderate values of $x\Delta$ and large values of $d\Delta$ the difference is insignificant.

Since the Fourier transform of a rectangular pulse is of $\sin(x)/x$ form, Equations 5.25 and 5.27 confirm the poor resolution performance predicted by the uncertainty function. If the signal at the input of the Fourier transform calculator had been multiplied by a time weighting function $m(t)$, Equation 5.25 would have become

$$C(j\omega) = \tfrac{1}{2}\,\mathscr{F}\{m(t)\,e(t)\,e(t+x-t_v)\} \tag{5.29}$$

One is now at liberty to choose a suitable $m(t)$ such that the above Fourier transform has low level sidelobes. Fortunately there are many well known functions which have this property—a good treatment is given by Blackman and Tukey [15]. One suitable function is

$$\left. \begin{array}{ll} m(t) = \dfrac{1+\alpha}{2} + \dfrac{1-\alpha}{2}\,\cos\left[\dfrac{2\pi}{d}\,(t-d/2)\right], & 0 < t < d \\[2em] m(t) = 0, & \begin{array}{l} t < 0 \\ t > d \end{array} \end{array} \right\} \tag{5.30}$$

that is

$$m(t) = \quad\begin{array}{c}\text{------1}\\ \text{--}\alpha\end{array}$$

A suitable value for α is 0.08 which results in the Hamming function [15]; the sidelobe performance is shown in Fig. 5.4. Strictly $m(t)$

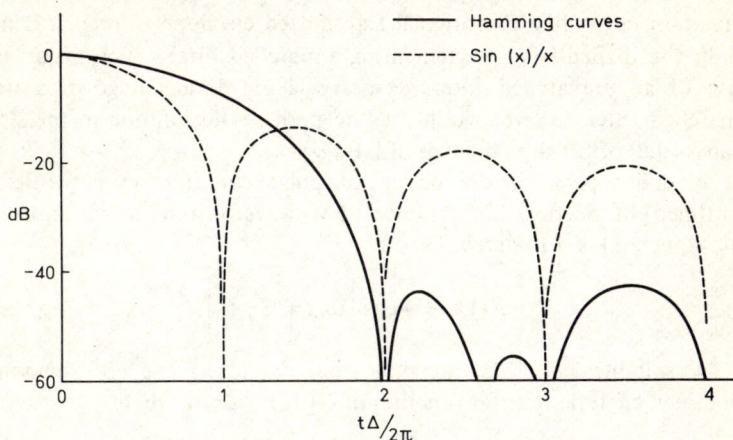

Fig. 5.4. The inverse F.T. of the Hamming function of Equation 5.36

should be varied according to the target delay to obtain the optimum shape (i.e. the function $m(t + x - t_v)$ should be used) but for large values of $d\Delta$ the reasoning which predicted no significant change for the unweighted receiver also applies to the weighted receiver.

5.2.2 The matched filter receiver

It can be seen that the manufacture and weighting of a Fourier transform receiver is relatively straightforward. Assuming that broadband mixers are available, the main problem is the manufacture of a 'real time' spectrum analyser or having to accept the slow readout of a conventional one. By contrast, the matched filter receiver is more difficult to implement. However, once the filtering problems have been overcome the output is inherently 'real time'.

It is virtually impossible to manufacture an exact matched filter for the rectangular pulse with linear FM. The complex analytic form of the required matched filter transfer function is

$$G^*[j(\omega - \omega_0)]$$

where $G(j\omega)$ is given by Equation A3.6 of Appendix 3. The requirement is for a filter with a baseband response

$$e^{j(\omega^2 d/2\Delta)}$$

cascaded with a complicated Fresnel integral function.

Matched filter receivers for linear FM employ unmatched filters called pulse compression networks. The term 'pulse compression' arises because the final decision making envelope is short, like the uncertainty function cut, while the original transmitted envelope is long. It is not only the difficulty of implementing a matched filter which makes the use of an unmatched filter necessary—a big disadvantage of a true matched filter receiver would be the poor resolution due to the slow rate of fall off of the sidelobes of $|\chi(\tau, o)|$.

A philosophy for the design of pulse compression networks is outlined in Section 7.7. A linear FM waveform with an arbitrary envelope $e(t)$ is considered, i.e.

$$f(t) = e(t) \cos\left[\omega_0 t + \frac{\Delta}{2d} t^2\right] \qquad (5.31)$$

A suitable pulse compression filter would have a corresponding complex analytic transfer function of $G[j(\omega - \omega_0)]$ where

$$G(j\omega) = B(j\omega)\, e^{j(\omega^2 d/2\Delta)} \qquad (5.32)$$

Note that the exponential term in Equation 5.32 is the same as that required for the matched filter. Group delay [11] is defined as $-(d\phi/d\omega)$ hence the above exponential term corresponds to a dispersive line having a delay proportional to frequency.

Section 7.7 (Equation 7.52) shows that if a signal of the form of $f(t)$, but having a Doppler offset of ω_d, is applied to such a pulse compression network, the RF envelope of the output will be given by

$$|h_d(t)| = \frac{d}{4\pi} \sqrt{\frac{2\pi}{d\Delta}} \left| \int_{-\infty}^{\infty} E(jx)b\left[\frac{dx}{\Delta} + \frac{d\omega_d}{\Delta} + t\right] e^{-jd(\omega_d + x)^2/(2\Delta)}\, dx \right| \qquad (5.33)$$

For a given transmitted envelope $e(t)$, $E(j\omega)$ can easily be calculated, thus to choose a suitable $B(j\omega)$ it is necessary to find a $b(t)$ which, when substituted in Equation 5.33 leads to a low sidelobe level.

The above procedure is a daunting task since in general Equation 5.33 can only be evaluated numerically. However, for very large values of dispersion factor it turns out that the term

$$\exp\left[-j\frac{d}{2\Delta}(\omega_d + x)^2\right] = \exp\left[-j\pi\left(\frac{2\pi}{d\Delta}\right)\left(\frac{d\omega_d}{2\pi} + \frac{dx}{2\pi}\right)^2\right]$$

can be considered equal to unity over the range of the other two functions in Equation 5.33. Section 7.7.3 shows that if this is true a substantial simplification results; Equation 5.33 becomes

$$|h_a(t)| = \frac{d}{2}\sqrt{\frac{2\pi}{d\Delta}}\left|m\left(t + \frac{d\omega_d}{\Delta}\right)\right| \tag{5.34}$$

where

$$m(t) = \mathscr{F}^{-1}\left[B(j\omega)\,e\left(\frac{d\omega}{\Delta}\right)\right] \tag{5.35}$$

Note that Equation 5.34 indicates complete delay-Doppler coupling, the effect of a Doppler offset being merely a shift in time without any change in shape.

In a practical situation Equations 5.34 and 5.35 could be used to make preliminary investigations of possible choices of $B(j\omega)$, a numerical evaluation of Equation 5.33 could then be carried out for specific values of $d\Delta$.

As in the case of time weighting, the Hamming function would be a good choice for $B(j\omega)$, the form would be

$$\left.\begin{aligned}
B(j\omega) &= \frac{1+\alpha}{2} + \frac{(1-\alpha)}{2}\cos\left[\frac{2\pi\omega}{\Delta}\right], & |\omega| &< \frac{\Delta}{2} \\
B(j\omega) &= 0, & |\omega| &> \frac{\Delta}{2}
\end{aligned}\right\} \tag{5.36}$$

where $\alpha = 0.08$. Hence

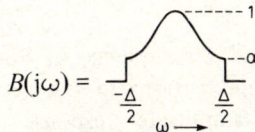

For a rectangular transmitted envelope let

Hence Equation 5.35 reduces to

$$m(t) = \mathscr{F}^{-1}\{B(j\omega)\} = \left[\frac{\alpha t^2\Delta^2 - 2\pi^2(1+\alpha)}{\pi t\,[t^2\Delta^2 - 4\pi^2]}\right]\sin\left(\frac{\Delta t}{2}\right) \tag{5.37}$$

Equation 5.37 is illustrated by Fig. 5.4–which also shows a $\sin(x)/x$ curve for comparison. Equation 5.34 predicts that the compressed pulse will have the shape of $m(t)$ for large values of dispersion factor; it can be seen that this waveform is a great improvement over the output from a matched filter since the sidelobes start at −42 dB rather than −13 dB. Figure 5.4 shows, however, that the main lobe is broader and that the 13 dB resolutions of the two waveforms are similar.

Figure 5.5 is the result of a numerical evaluation of Equation 5.33 for the Hamming function and a dispersion factor of 100. It can be seen that $d\Delta$ must be much larger than 100 for Equation 5.34 to be valid. The author's numerical integration program is limited by computer rounding errors to a maximum value of $d\Delta$ of 100.

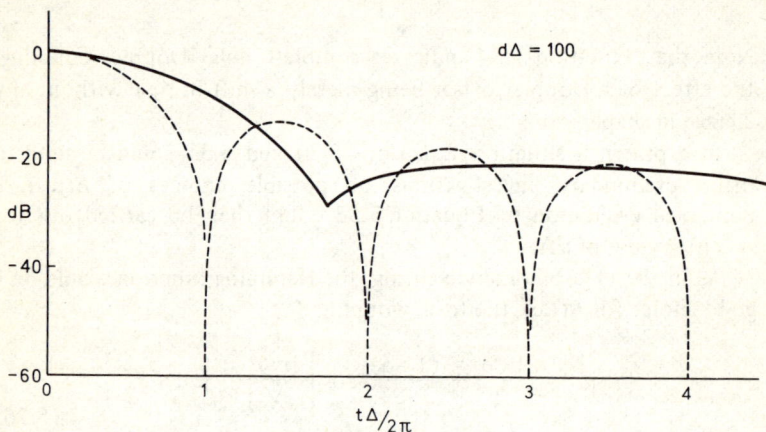

Fig. 5.5. *The result of using the Hamming function with a low dispersion factor*

An important practical consequence of Equation 5.35 is that, for large values of $d\Delta$, imperfections in the actual shape of the pulse compression network amplitude response can be corrected by modifications to the transmitted envelope. Drastic envelope modifications defeat the object of transmitting more energy in a given bandwidth, but small corrections to optimise the sidelobe level can probably be carried out more conveniently on the envelope than on the filter. It should be particularly noted that a rectangular envelope provides the sharp cut off required in the Hamming function frequency response.

5.2.3 The effect of duty cycle

One further point to be discussed—applicable to both types of receiver—is the question of the pulse repetition frequency period or, what is more relevant, duty cycle.

As in the case of the constant carrier frequency pulse of Section 5.1, the pulse repetition frequency period k should be chosen to eliminate second time round ambiguities. The problem of duty cycle did not have to be faced in Section 5.1 since, for good precision, short pulses had to be used and the duty cycle was inherently low—this in fact is the big disadvantage of the constant carrier frequency pulse.

If the linear FM waveform is used duty cycles can approach unity and still give good precision and resolution if the dispersion factor $d\Delta$ is large enough. The advantage of using duty cycles of much less than 0.5 is that the resultant uncertainty function has areas of no overlap, hence, no swamping can occur for targets separated by delays greater than $2d$. Duty cycles of greater than 0.5 lead to overlapping uncertainty functions (see Section 2.3); the sidelobes of the uncertainty function fall to a minimum value for τ equal to approximately $k/2$ and then begin to rise again, with no clear region.

It is common, in the established literature, to refer to FM systems having a duty cycle much less than 0.5 as 'chirp radars'. The term 'FM CW' is used to refer to FM radars having duty cycles approaching unity.

Whether the effect of overlap is important depends upon the dispersion factor used. It can be seen from Section 3.2 that $\chi(0, 0)$ is equal to d, and that the envelope of $|\chi(\tau, 0)|$ falls off according to the law

$$\frac{2d}{(d\Delta)(\tau/d)}$$

Thus at the critical value of $\tau/d = 0.5$ the envelope of $|\chi(\tau, 0)|$ is

$$20 \log_{10}\left(\frac{d\Delta}{4}\right)$$

dB down with respect to the amplitude of the main lobe.

5.2.4 The effect of high dispersion factor

A short discussion on the circumstances under which Equation 5.33 reduces to Equation 5.34 will now be given.

It follows from the definitions of d and Δ that, for large values of the dispersion factor, most of the transmitted energy will lie in a band roughly Δ rad/s wide; thus the bandwidth of a useful $B(j\omega)$ would have to be in the region of Δ to yield a good output signal to noise ratio.

$E(j\omega)$ and $b(t)$ for the extreme examples of a rectangular envelope and a rectangular $B(j\omega)$ are illustrated in Figs 5.6, 5.7 and 5.8.

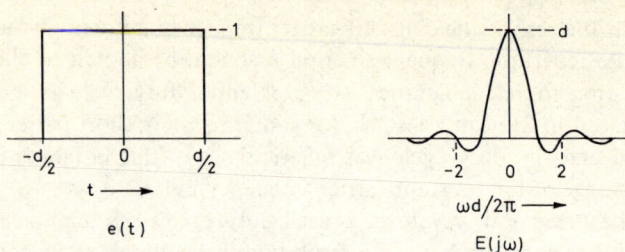

Fig. 5.6. A typical transmitted envelope and its spectrum

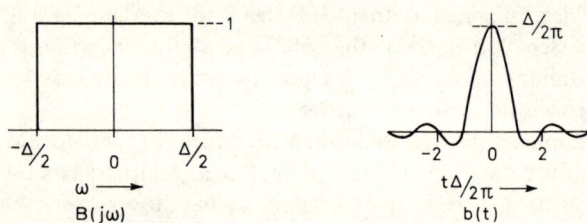

Fig. 5.7. A possible B(jω) and its impulse response

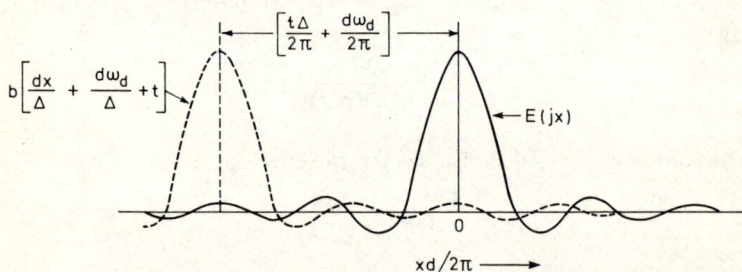

Fig. 5.8. The functions to be multiplied in Equation 5.33.

It is only necessary to carry out the integration in Equation 5.33 over those values of x for which the product

$$E(jx)b\left[\frac{dx}{\Delta} + \frac{d\omega_d}{\Delta} + t\right]$$

is significantly large. Figure 5.8 shows that, for moderate values of t and ω_d the product will be small for large values of $xd/2\pi$, hence it is possible to have a sufficiently large value of $d\Delta$ for the exponential term in Equation 5.33 to remain sensibly constant over the effective range of integration.

Chapter 6

LAPLACE AND FOURIER TRANSFORMS

Most electronics engineers are familiar with the one-sided Laplace transform which leads to the modern pole-zero method of circuit design. Signal theory is usually written in terms of the Fourier transform which can be regarded as a special case of the two-sided Laplace transform.

The two-sided Laplace transform is introduced in Section 6.1 and its relationship to the Fourier transform is explained in Section 6.4. It is suggested that the two-sided Laplace transform definition be accepted as a 'personal' definition even when one-sided Laplace transforms are being discussed. Two simple rules, given in Section 6.3, show the slight change of viewpoint necessary for the above to be successful.

A thorough treatise on the two-sided Laplace transform is contained in the excellent book by van der Pol and Bremmer [8].

The advantages of the two-sided Laplace transform may be summarised as follows:

(1) The class of functions suited to an operational treatment becomes much larger.
(2) The transformation rules are considerably simplified.
(3) The entire treatment becomes more rigorous than the usual presentation of the one-sided integral in technical books.

6.1 THE TWO-SIDED LAPLACE TRANSFORM

If

$$f(p) = p \int_{-\infty}^{\infty} e^{-pt} h(t) \, dt$$

converges for $\alpha < \text{Re}(p) < \beta$, then

$$h(t) = \frac{1}{2\pi j} \int_{c-j\infty}^{c+j\infty} e^{pt} \frac{f(p)}{p} \, dp$$

where $\alpha < c < \beta$

The infinite strip of the p plane bounded by $\alpha < \text{Re}(p) < \beta$ is termed the 'strip of convergence'. The precise meaning of convergence, and the necessary restrictions on $h(t)$ for the above to be true are set out in [8].

The above notation is known as the 'p multiplied' form of the Laplace transform. It was used to allow the results obtained to be more easily identified with the work of Heaviside (1850-1925). It is more usual to call the function $f(p)/p$ 'the Laplace transform'. It is also convenient to refer to $h(t)$ as a time function and $H(p)$ as its Laplace transform.

The Laplace transform of $f(t)$ is defined by

$$\mathcal{L}\{f(t)\} = F(p) = \int_{-\infty}^{\infty} f(t)\, e^{-pt}\, dt \tag{6.1}$$

If the integral converges for $\alpha < \mathrm{Re}(p) < \beta$ the inversion integral is

$$f(t) = \frac{1}{2\pi j} \int_{c-j\infty}^{c+j\infty} F(p)\, e^{pt}\, dp \tag{6.2}$$

where $\alpha < c < \beta$.

For all practical purposes the above formulae define unique functions. Thus, if two functions have equal Laplace transforms (regarding the strip of convergence as part of the transform definition) then they are equivalent.

6.2 THE SIGNIFICANCE OF THE STRIP OF CONVERGENCE

A specification of a two-sided Laplace transform is not complete without reference to its strip of convergence. Although the limits of the required strip of convergence can be formally calculated they usually become apparent as the Laplace transform is evaluated.

The above remarks, together with the effect of choosing the wrong strip of convergence can be illustrated by a simple example. Consider

$$u(t) = 1, \qquad t > 0$$
$$u(t) = 0.5, \qquad t = 0$$
$$u(t) = 0, \qquad t < 0$$

Applying Equation 6.1 leads to

$$U(p) = \int_{-\infty}^{\infty} u(t)\, e^{-pt}\, dt = \int_{0}^{\infty} e^{-pt}\, dt = \left[\frac{e^{-pt}}{-p}\right]_{0}^{\infty}$$

The upper limit gives an infinite value unless $\mathrm{Re}(p) > 0$. Hence

$$\mathcal{L}\{u(t)\} = \frac{1}{p}, \qquad \mathrm{Re}(p) > 0$$

Contrast the case for $u(-t)$, i.e.

$$u(-t) = 0, \qquad t > 0$$

$$u(-t) = 0.5, \qquad t = 0$$

$$u(-t) = 1, \qquad t < 0$$

$$\mathcal{L}\{u(-t)\} = \int_{-\infty}^{\infty} u(-t)\, e^{-pt}\, dt = \int_{-\infty}^{0} e^{-pt}\, dt = \left[\frac{e^{-pt}}{-p}\right]_{-\infty}^{0}$$

This time the lower limit will give an infinite value unless $\text{Re}(p) < 0$. Hence

$$\mathcal{L}\{u(-t)\} = -\frac{1}{p}, \qquad \text{Re}(p) < 0$$

Thus the function whose Laplace transform is $1/p$ could either mean $u(t)$ or $-u(-t)$. Specification of the strip of convergence removes the ambiguity.

It can be shown that all one-sided functions (i.e. functions which are zero for $t < 0$) lead to a strip of convergence which lies to the right of all the poles of the resulting Laplace transform.

6.3 THE ONE-SIDED LAPLACE TRANSFORM

It is more usual to define the Laplace transform of $f(t)$ as

$$\int_{0}^{\infty} f(t)\, e^{-pt}\, dt.$$

The expression, having a lower limit of zero, is known as the 'one-sided' Laplace transform since values of $f(t)$ for $t < 0$ are ignored.

It is perfectly feasible to always regard the Laplace transform in terms of the two-sided definition of Section 6.1, and read work based upon the one-sided definition as if it had been written in terms of two-sided transforms. It is only necessary to remember the following two points:

(1) References to $f(t)$ in the one-sided treatment mean $f(t)u(t)$, where $u(t)$ is the unit step function.
(2) The strip of convergence of the Laplace transform of $f(t)u(t)$ is that region of the p plane which lies to the right of all the Laplace transform poles.

6.4 THE FOURIER TRANSFORM

The Fourier transform of $f(t)$ is defined as

$$\mathcal{F}\{f(t)\} = \int_{-\infty}^{\infty} f(t)\, e^{-j\omega t}\, dt \qquad (6.3)$$

If the integral converges, the inversion integral is given by

$$f(t) = \frac{1}{2\pi} \int_{-\infty}^{\infty} \mathcal{F}\{f(t)\}\, e^{j\omega t}\, d\omega \qquad (6.4)$$

To see the relationship between Equations 6.3 and 6.4 and Equations 6.1 and 6.2 (defining the two-sided Laplace transform) it is necessary to remember that p is a complex variable which may be written in the form

$$p = c + j\omega$$

where c and ω are real variables. If the definition integral of the Laplace transform (Equation 6.1) converges for $\mathrm{Re}(p) = 0$, it is permissible to put $p = j\omega$ and $c = 0$ in the Laplace transform inversion integral. The (two-sided) Laplace transform then reduces to the Fourier transform and one may write

$$\mathcal{F}\{f(t)\} = F(j\omega)$$

If the Laplace transform of $f(t)$ does not converge for $\mathrm{Re}(p) = 0$, the Fourier transform does not exist. Thus the Laplace transform definition of Section 6.1 embraces a larger class of functions than does that of the Fourier transform.

Some examples are

(1) $f(t) = e^{-at}u(t)$. Therefore

$$F(p) = \frac{1}{p+a}, \qquad \mathrm{Re}(p) > -a$$

Hence

$$\mathcal{F}\{f(t)\} = F(j\omega) = \frac{1}{a+j\omega}, \qquad \text{for } a > 0$$

(2) $f(t) = e^{at}u(t)$. Therefore

$$F(p) = \frac{1}{p-a}, \qquad \mathrm{Re}(p) > a$$

Hence $\mathcal{F}\{f(t)\}$ does not exist, for $a > 0$.

(3)

$$f(t) = e^{-at}, \qquad t > 0$$
$$f(t) = e^{bt}, \qquad t < 0$$

Therefore

$$F(p) = \frac{1}{p+a} + \frac{1}{b-p}, \qquad -a < \text{Re}(p) < b$$

Hence, for the Laplace transform to exist it is necessary to have $-a < b$. For the Fourier transform to exist it is necessary to have $-a < 0 < b$.

6.5 THE PHYSICAL INTERPRETATION OF LAPLACE AND FOURIER TRANSFORMS

It is possible to visualise both the Laplace and Fourier transforms in terms of frequency spectra. Equation 6.4 can be regarded as expressing $f(t)$ as an infinite sum of time functions of the form $e^{j\omega t}$. The term '$d\omega$' ensures that the contribution at any one frequency is vanishingly small, and the multiplier $F(j\omega)/2\pi$ may be regarded as the spectral density, measured in volts per rad/s. Hence the units of $F(j\omega)$ are volts/Hz.

The function $e^{j\omega t}$ is a complex quantity, i.e.

$$e^{j\omega t} = \cos(\omega t) + j \sin(\omega t)$$

which can be visualised as a rotating vector. For +ve frequencies $\omega > 0$ and the rotation is anticlockwise, while for −ve frequencies $\omega < 0$ and the rotation is clockwise. For real functions of time the Fourier transform symmetries (Section 6.6) are such that terms in $e^{j\omega t}$ combine with terms in $e^{-j\omega t}$ to give real sinusoids. The interpretation carries directly over to the Laplace transform, the only difference being that the basic time function becomes

$$e^{pt} = e^{ct} e^{j\omega t}$$

which can be regarded as an expanding rotating vector ($c > 0$) or a contracting rotating vector ($c < 0$). Thus the physical interpretation of Equation 6.2 is that $f(t)$ can be regarded as consisting of any one of a number of spectra of the form $e^{ct} e^{j\omega t}$; c being in the range $\alpha < c < \beta$. The spectral density is given by the expression $F(p)/2\pi j$.

It should be noted that it is not essential to have a physical explanation for a mathematical result; indeed some authors scorn a physical interpretation of the Laplace transform. The big advantage of physical interpretations is that they allow reasoning by analogy—there is nothing wrong with this as long as the results obtained are subsequently verified by rigorous methods. In many cases the most difficult part of a mathematical proof is deciding what has to be proved in the first place!

6.6 FOURIER TRANSFORM SYMMETRIES

This section will be used to prove the following results:

If
$$\mathscr{F}\{f(t)\} = F(j\omega)$$

$$\mathscr{F}\{F(jt)\} = 2\pi f(-\omega) \tag{6.5}$$

$$\mathscr{F}\{f(at)\} = \frac{1}{|a|}\, F\left(j\frac{\omega}{a}\right) \tag{6.6}$$

Also that the following necessary and sufficient conditions apply

$$F(-j\omega) = F^*(j\omega), \qquad \text{for } f(t) \text{ real} \tag{6.7}$$

$$F(-j\omega) = -F^*(j\omega), \quad \text{for } f(t) \text{ imaginary} \tag{6.8}$$

$$f(-t) = f^*(t), \qquad \text{for } F(j\omega) \text{ real} \tag{6.9}$$

$$f(-t) = -f^*(t), \qquad \text{for } F(j\omega) \text{ imaginary} \tag{6.10}$$

Other results are stated in Section 6.10.

6.6.1 Proof of Equation 6.5

Using Equation 6.3 and remembering that the integration variable t is a dummy variable, gives

$$\mathscr{F}\{F(jt)\} = \int_{-\infty}^{\infty} F(jx)\, e^{-j\omega x}\, dx$$

Similarly, from

$$f(t) = \frac{1}{2\pi} \int_{-\infty}^{\infty} F(jx)\, e^{j\omega x}\, dx$$

Hence
$$\mathscr{F}\{F(jt)\} = 2\pi f(-\omega).$$

EXAMPLE

Since

$$\mathscr{F}\left\{ \underset{\substack{-a \quad a \\ t \to}}{\boxed{}}^{\,\cdots 1} \right\} = \frac{2}{\omega}\, \sin(a\omega) \tag{6.11}$$

then

$$\mathscr{F}\left\{ \frac{2}{t}\, \sin(at) \right\} = 2\pi \left[\underset{\substack{-a \quad a \\ -\omega \to}}{\boxed{}}^{\,\cdots 1} \right] \tag{6.12}$$

6.6.2 Proof of Equation 6.6

$$\mathcal{F}\{f(at)\} = \int_{-\infty}^{\infty} f(at)\,e^{-j\omega t}\,dt$$

Changing the dummy variable to t/a gives

$$\mathcal{F}\{f(at)\} = \int_{-\infty}^{\infty} f(t)\,e^{-j\omega t/a}(1/a)\,dt, \qquad a > 0$$

$$\mathcal{F}\{f(at)\} = -\int_{-\infty}^{\infty} f(t)\,e^{-j\omega t/a}(1/a)\,dt, \qquad a < 0$$

which is expressed compactly by Equation 6.6.

A frequently used application of Equation 6.6 is the case resulting from $a = -1$, i.e.

$$\mathcal{F}\{f(-t)\} = F(-j\omega) \tag{6.13}$$

6.6.3 Proof of Equations 6.7 to 6.10

The proof of Equations 6.7 to 6.10 follows from the formulae for the real and imaginary parts of a general function $g(z)$

$$\text{Re}\{g(z)\} = \frac{1}{2}\,[g(z) + g^*(z)] \tag{6.14}$$

$$\text{Im}\{g(z)\} = \frac{1}{2j}\,[g(z) - g^*(z)] \tag{6.15}$$

Thus necessary and sufficient conditions are

$$g(z) = g^*(z), \qquad \text{for } g(z) \text{ real}$$

$$g(z) = -g^*(z), \qquad \text{for } g(z) \text{ imaginary}$$

It is shown in Section 6.7 that

$$\mathcal{F}\{f^*(t)\} = F^*(-j\omega)$$

$$\mathcal{F}^{-1}\{F^*(j\omega)\} = f^*(-t)$$

Hence Equations 6.7 to 6.10 follow.

6.7 THE FOURIER TRANSFORM OF A CONJUGATE FUNCTION

The following results will be proved

$$\mathscr{F}\{f^*(t)\} = F^*(-j\omega) \qquad (6.16)$$

$$\mathscr{F}^{-1}\{[F^*(j\omega)]\} = f^*(-t) \qquad (6.17)$$

The definition integral (Section 6.4) gives

$$F(-j\omega) = \int_{-\infty}^{\infty} f(t)\, e^{j\omega t}\, dt$$

Hence

$$F^*(-j\omega) = \int_{-\infty}^{\infty} f^*(t)\, e^{-j\omega t}\, dt$$

which proves Equation 6.16.

Since Equation 6.16 is true, Equation 6.17 follows from Equation 6.13.

6.8 LIMITING CASES OF THE FOURIER TRANSFORM

Certain useful functions do not, strictly speaking, have Fourier transforms. They can however be regarded as limiting cases of functions which do. Consider, for example, $u(t)$, the unit step function discussed in Section 6.2.

$$\mathscr{L}\{u(t)\} = \frac{1}{p}, \qquad \mathrm{Re}(p) > 0$$

Since the strip of convergence does not include $\mathrm{Re}(p) = 0$ the Fourier transform of the unit step does not exist.

If, for the purpose of analysis, one is prepared to accept that a decaying function with a 'half-life' of millions of years will have the same effect as a unit step, it is permissible to write

$$\mathscr{F}\{u(t)\} = \underset{\epsilon \to +0}{\mathrm{Lt}}\ \mathscr{F}\{e^{-\epsilon t} u(t)\}$$

Since

$$\mathscr{L}\{e^{-\epsilon t} u(t)\} = \frac{1}{p+\epsilon}, \qquad \mathrm{Re}(p) > -\epsilon$$

it follows that

$$\mathscr{F}\{u(t)\} = \underset{\epsilon \to +0}{\mathrm{Lt}} \left\{ \frac{1}{\epsilon + j\omega} \right\} = \underset{\epsilon \to +0}{\mathrm{Lt}} \left\{ \frac{\epsilon}{\epsilon^2 + \omega^2} - j\frac{\omega}{\epsilon^2 + \omega^2} \right\}$$

The first expression in the right-hand limit will be recognised as $\pi\delta(\omega)$ (see Section 6.9) while the second expression becomes $-j/\omega$ for $\omega \neq 0$, and zero for $\omega = 0$. Hence

$$\mathscr{F}\{u(t)\} = \pi\delta(\omega) + i(\omega) \tag{6.18}$$

where

$$\left.\begin{array}{ll} i(\omega) = \dfrac{1}{j\omega}, & \omega \neq 0 \\[2mm] i(\omega) = 0, & \omega = 0 \end{array}\right\} \tag{6.19}$$

Similarly, for a reversed unit step

$$\mathscr{F}\{u(-t)\} = \pi\delta(\omega) - i(\omega) \tag{6.20}$$

Combining Equations 6.18 and 6.20 to give the Fourier transform of unity

$$\mathscr{F}\{1\} = \mathscr{F}\{u(t) + u(-t)\} = 2\pi\delta(\omega) \tag{6.21}$$

Alternatively

$$\mathscr{L}\{1\} = \underset{\epsilon \to +0}{\text{Lt}}\ \mathscr{L}\{e^{-\epsilon|t|}\}$$

$$\mathscr{L}\{e^{-\epsilon|t|}\} = \frac{2\epsilon}{\epsilon^2 - p^2}, \qquad -\epsilon < \text{Re}(p) < \epsilon$$

Therefore

$$\mathscr{F}\{1\} = \underset{\epsilon \to +0}{\text{Lt}}\ \left\{\frac{2\epsilon}{\epsilon^2 + \omega^2}\right\} = 2\pi\delta(\omega) \tag{6.22}$$

The above results are consistent with the physical interpretation discussed in Section 6.5 where $F(j\omega)/2\pi$ had the significance of spectral density.

Equation 6.18 credits the unit step with a spectral density of 0.5, concentrated at d.c., plus some high frequency components due to the sharp transition at $t = 0$. Equation 6.22 shows the unit d.c. level to have a spectral density of 1, concentrated at d.c., and no high frequency components.

The precise behaviour of $\mathscr{F}\{u(t)\}$ at $\omega = 0$, given by Equations 6.18 and 6.19, is used in Section 7.2; it also allows a quoted result for the signum function to be verified. Consider

$$\text{sgn}(t) = \begin{cases} 1, & t > 0 \\ 0, & t = 0 \\ -1, & t < 0 \end{cases}$$

Hence

$$\mathrm{sgn}(t) = 2u(t) - 1$$

Using Equations 6.18 and 6.22

$$\mathcal{F}\{\mathrm{sgn}(t)\} = 2i(\omega) = \left.\begin{array}{ll} 2/j\omega, & \omega \neq 0 \\ 0, & \omega = 0 \end{array}\right\} \qquad (6.23)$$

which, together with the remarks on notation in Section 6.11, verifies a result given by van der Pol and Bremmer [8] (page 114, Equation 54).

The Fourier transform symmetry equation (Equation 6.5) can be used with Equation 6.18 to prove another result which is required for Section 7.2.

Since

$$\mathcal{F}\{u(t)\} = \pi\delta(\omega) + i(\omega)$$

it follows that

$$\mathcal{F}\{\pi\delta(t) + i(t)\} = 2\pi u(-\omega)$$

Hence

$$\mathcal{F}^{-1}\{u(\omega)\} = \tfrac{1}{2}\delta(t) - \frac{1}{2\pi}\, i(t) \qquad (6.24)$$

where

$$i(t) = \left.\begin{array}{ll} \dfrac{1}{jt}, & t \neq 0 \\[2mm] 0, & t = 0 \end{array}\right\} \qquad (6.25)$$

6.9 THE DELTA FUNCTION

This section will be used to give a short description of the delta function, $\delta(t)$, and some of its properties. A full treatment will be found in van der Pol and Bremmer [8].

The delta function is defined through the following properties

$$\delta(x) = \left\{\begin{array}{ll} 0, & x \neq 0 \\ \infty, & x = 0 \end{array}\right\} \qquad (6.26)$$

$$\int_{-\infty}^{\infty} \delta(x)\, dx = 1 \qquad (6.27)$$

Although Equations 6.26 and 6.27 do not define a function in the ordinary mathematical sense, the delta function does have a rigorous mathematical basis [19].

It is in order, and perhaps more satisfactory, to regard the delta function as a 'short-hand' notation for a limit, involving a normal function, which possesses the above properties if the limit is taken as the last operation. As an example of such a function, consider the integral

$$\int \frac{dx}{a^2 + x^2} = \frac{1}{a} \tan^{-1}\left(\frac{x}{a}\right)$$

where the principal value of $\tan^{-1}$ is intended [10]. Hence

$$\int_{-\infty}^{\infty} \frac{a\,dx}{a^2 + x^2} = \left[\frac{\pi}{2} + \frac{\pi}{2}\right] = \pi$$

The above results lead to the following representation of the delta function

$$\delta(x) = \underset{\epsilon \to +0}{\text{Lt}} \left\{ \frac{\epsilon}{\pi(x^2 + \epsilon^2)} \right\} = \underset{\lambda \to \infty}{\text{Lt}} \left\{ \frac{\lambda}{\pi(\lambda^2 x^2 + 1)} \right\} \qquad (6.28)$$

A similar treatment allows trains of delta functions to be represented as limits. As an example

$$\sum_{n=-\infty}^{\infty} \delta(t - 2\pi n) = \underset{r \to 1-0}{\text{Lt}} \left\{ \frac{1 - r^2}{2\pi(1 - 2r \cos t + r^2)} \right\} \qquad (6.29)$$

The most useful characteristic of the delta function is its sifting property,

$$\int_{-\infty}^{\infty} h(x)\,\delta(t - x)\,dx = h(t) \qquad (6.30)$$

Equation 6.30 follows from Equation 6.27, remembering that $\delta(t - x)$ is zero except at the point $x = t$.

An alternative representation of the delta function can be obtained by considering its Fourier transform.

$$\mathscr{F}\{\delta(t)\} = \int_{-\infty}^{\infty} \delta(t)\,e^{-j\omega t}\,dt = 1 \qquad (6.31)$$

The result of Equation 6.31 follows from Equation 6.30 by noting that

$$e^{-j\omega t} = 1, \qquad \text{for } t = 0$$

Equation 6.31 can also be written in the form of an inverse Fourier transform giving

$$\delta(t) = \frac{1}{2\pi} \int_{-\infty}^{\infty} e^{j\omega t}\,d\omega \qquad (6.32)$$

Equation 6.32 may be used to obtain another useful result—the interpretation of $\delta(at + b)$.

$$\delta(at + b) = \frac{1}{2\pi} \int_{-\infty}^{\infty} e^{j(at+b)\omega} \, d\omega$$

Changing the dummy variable to ω/a gives

$$\delta(at + b) = \frac{1}{2\pi a} \int_{-\infty}^{\infty} e^{j(t + b/a)\omega} \, d\omega, \qquad a > 0$$

$$\delta(at + b) = \frac{-1}{2\pi a} \int_{-\infty}^{\infty} e^{j(t + b/a)\omega} \, d\omega, \qquad a < 0$$

Hence

$$\delta(at + b) = \frac{1}{|a|} \delta \left(t + \frac{b}{a} \right) \tag{6.33}$$

Failure to appreciate the result given by Equation 6.33 once led to the discussion of a 'paradox' in the correspondence section of *Proc. I.R.E.* (November, 1961; February, 1962).

6.9.1 Derivation of the Fourier series

The delta function is used by recognising the significance of limits such as Equations 6.28 and 6.29 and using properties such as Equations 6.30 to 6.33. A good example of this procedure is given by a derivation of the Fourier series from a consideration of the two-sided Laplace transform of a repetitive function.

Consider the function

$$f_R(t) = \sum_{n=-\infty}^{\infty} e^{-|n|ck} f(t - nk) \tag{6.34}$$

where n takes on integer values (including zero), c, k are real +ve constants and $\mathcal{L}\{f(t)\}$ converges for $-\alpha < \mathrm{Re}(p) < \infty$, with $\alpha > c$. Equation 6.34 may be written in the form

$$f_R(t) = f_1(t) + f_2(t) - f(t) \tag{6.35}$$

where

$$f_1(t) = f(t) + e^{-ck} f(t - k) + e^{-2ck} f(t - 2k) + \ldots$$

$$f_2(t) = f(t) + e^{-ck} f(t + k) + e^{-2ck} f(t + 2k) + \ldots$$

It will be noted that the factor $e^{-|n|ck}$ ensures that the 'pulses' decrease in amplitude each side of the point $t = 0$. This effect can be made negligible by allowing c to tend to zero.

From Equation 6.35

$$F_1(p) = F(p)[1 + e^{-ck}e^{-pk} + e^{-2ck}e^{-2pk} + \ldots] \qquad (6.36)$$

$$F_2(p) = F(p)[1 + e^{-ck}e^{pk} + e^{-2ck}e^{2pk} + \ldots] \qquad (6.37)$$

Equations 6.36 and 6.37 are geometric progressions which summed to infinity give

$$F_1(p) = \frac{F(p)}{1 - e^{-ck}e^{-pk}}, \qquad |e^{-ck}e^{-pk}| < 1 \qquad (6.38)$$

$$F_2(p) = \frac{F(p)}{1 - e^{-ck}e^{pk}}, \qquad |e^{-ck}e^{pk}| < 1 \qquad (6.39)$$

The convergence conditions of Equations 6.38 and 6.39 reduce to $\text{Re}(p) > -c$ and $\text{Re}(p) < c$, respectively. Hence

$$F_R(p) = F(p)\left[\frac{1}{1 - e^{-ck}e^{-pk}} + \frac{1}{1 - e^{-ck}e^{pk}} - 1\right] \qquad (6.40)$$

where $-c < \text{Re}(p) < c$. Since the strip of convergence includes $\text{Re}(p) = 0$, $f_R(t)$ has a Fourier transform given by

$$F_R(j\omega) = F(j\omega)\left[\frac{1}{1 - e^{-ck}e^{-j\omega k}} + \frac{1}{1 - e^{-ck}e^{j\omega k}} - 1\right] \qquad (6.41)$$

Equation 6.41 simplifies to

$$F_R(j\omega) = F(j\omega)\left[\frac{1 - e^{-2ck}}{1 - 2e^{-ck}\cos\omega k + e^{-2ck}}\right] \qquad (6.42)$$

Putting $e^{-ck} = r$ it is seen that for $c \to +0$, $r \to 1 - 0$. If Equations 6.42 and 6.29 are compared, it follows that

$$\underset{c \to +0}{\text{Lt}}\{F_R(j\omega)\} = 2\pi F(j\omega)\sum_{n=-\infty}^{\infty}\delta(\omega k - 2\pi n) \qquad (6.43)$$

The Fourier transform inversion integral gives

$$f_R(t) = \frac{1}{2\pi}\int_{-\infty}^{\infty}F_R(j\omega)\,e^{j\omega t}\,d\omega \qquad (6.44)$$

Substituting Equation 6.43 in 6.44 and using Equation 6.33 gives

$$\underset{c \to +0}{\text{Lt}}\{f_R(t)\} = \frac{1}{k}\int_{-\infty}^{\infty}F(j\omega)\,e^{j\omega t}\sum_{n=-\infty}^{\infty}\delta\left(\omega - \frac{2\pi n}{k}\right)d\omega \qquad (6.45)$$

Table 6.1

Property	Result		
$\mathscr{F}\{f(t)\}$	$F(j\omega)$	$F_1(\omega)$	$F_2(f)$
$\mathscr{L}\{f(t)\}$	$F(p)$†	$F_1(-jp)$†	$F_2\left(-j\dfrac{p}{2\pi}\right)$†
$\mathscr{F}\{f^*(t)\}$	$F^*(-j\omega)$	$F_1^*(-\omega)$	$F_2^*(-f)$
$\mathscr{F}\{f(at)\}$	$\dfrac{1}{|a|}F\left(j\dfrac{\omega}{a}\right)$	$\dfrac{1}{|a|}F_1\left(\dfrac{\omega}{a}\right)$	$\dfrac{1}{|a|}F_2\left(\dfrac{f}{a}\right)$
$\mathscr{F}\{f(t-\tau)\}$	$e^{-j\omega\tau}F(j\omega)$	$e^{-j\omega\tau}F_1(\omega)$	$e^{-j2\pi f\tau}F_2(f)$
$\mathscr{F}\{f(t)\,e^{j\omega_0 t}\}$	$F[j(\omega-\omega_0)]$	$F_1(\omega-\omega_0)$	$F_2\left(f-\dfrac{\omega_0}{2\pi}\right)$
$\mathscr{F}\{f(t)g(t)\}$	$\dfrac{1}{2\pi}\displaystyle\int_{-\infty}^{\infty}F(jx)G[j(\omega-x)]\,dx$	$\dfrac{1}{2\pi}\displaystyle\int_{-\infty}^{\infty}F_1(x)G_1(\omega-x)\,dx$	$\displaystyle\int_{-\infty}^{\infty}F_2(x)G_2(f-x)\,dx$
$\displaystyle\int_{-\infty}^{\infty}f(x)g(t-x)\,dx$	$\mathscr{F}^{-1}\{F(j\omega)G(j\omega)\}$	$\mathscr{F}^{-1}\{F_1(\omega)G_1(\omega)\}$	$\mathscr{F}^{-1}\{F_2(f)G_2(f)\}$
$f(t)$	$\dfrac{1}{2\pi}\displaystyle\int_{-\infty}^{\infty}F(j\omega)\,e^{j\omega t}\,d\omega$	$\dfrac{1}{2\pi}\displaystyle\int_{-\infty}^{\infty}F_1(\omega)\,e^{j\omega t}\,d\omega$	$\displaystyle\int_{-\infty}^{\infty}F_2(f)\,e^{j2\pi ft}\,df$
$\mathscr{F}^{-1}\{f(f)\}=\mathscr{F}^{-1}\left\{f\left(\dfrac{\omega}{2\pi}\right)\right\}$	$F(-j2\pi t)$	$F_1(-2\pi t)$	$F_2(-t)$

† The expression formed from the Fourier transform in this way does not necessarily represent the Laplace transform throughout the strip of convergence.

Applying the result of Equation 6.30 to 6.45 leads to the Fourier series

$$\operatorname*{Lt}_{c \to +0} \{f_R(t)\} = \frac{1}{k} \sum_{n=-\infty}^{\infty} F\left[j\frac{2\pi n}{k}\right] e^{j(2\pi n/k)t} \qquad (6.46)$$

6.10 FOURIER TRANSFORM NOTATION

The expression defined as the Fourier transform in Section 6.4 can be called $F(j\omega)$, $F(\omega)$ or $F(f)$ depending upon the author. Although the differences are mathematically trivial they can lead to confusion when theorems are quoted.

For the convenience of the reader, various results are quoted in Table 6.1 in the three 'languages'.

DEFINITION

$$\mathscr{F}\{f(t)\} = F(j\omega) = F_1(\omega) = F_2(f) = \int_{-\infty}^{\infty} f(t)\, e^{-j\omega t}\, dt$$

where

$$\omega = 2\pi f.$$

6.11 USING THE *P*-MULTIPLIED LAPLACE TRANSFORM NOTATION

Van der Pol and Bremmer [8] have given extensive lists of Laplace transform theorems and results under the respective headings of 'grammar' and 'dictionary'. As the information is given in the *p*-multiplied form it is useful to be able to translate it into the standard notation.

A typical 'dictionary' entry is

$$u(t) \to 1, \quad \text{valid for } 0 < \operatorname{Re}(p) < \infty$$

To convert this to the standard notation, simply divide the given Laplace transform by p. Thus

$$\mathscr{L}\{u(t)\} = \frac{1}{p}, \quad \text{valid for } 0 < \operatorname{Re}(p) < \infty$$

The change in notation does not affect the strip of convergence.

The procedure in the case of theorems is slightly more complicated. Some results in the p-multiplied notation are

If
$$h(t) \rightarrow f(p), \quad \text{valid for } \alpha < \text{Re}(p) < \beta$$

Then
$$h(t+\lambda) \rightarrow e^{\lambda p} f(p), \quad \text{valid for } \alpha < \text{Re}(p) < \beta$$

Also
$$e^{-\lambda t} h(t) \rightarrow \frac{p}{p+\lambda} f(p+\lambda), \quad \text{valid for } \alpha - \text{Re}(\lambda) < \text{Re}(p) < \beta - \text{Re}(\lambda)$$

To convert to the standard notation

(1) Divide the right-hand expression by p.
(2) Replace $f(z)$ by $zf(z)$.

The above examples then become, if
$$\mathcal{L}\{h(t)\} = f(p), \quad \text{valid for } \alpha < \text{Re}(p) < \beta$$

then
$$\mathcal{L}\{h(t+\lambda)\} = e^{\lambda p} f(p), \quad \text{valid for } \alpha < \text{Re}(p) < \beta$$

Also
$$\mathcal{L}\{e^{-\lambda t} h(t)\} = f(p+\lambda), \quad \text{valid for } \alpha - \text{Re}(\lambda) < \text{Re}(p) < \beta - \text{Re}(\lambda).$$

6.12 THE DISCRETE FOURIER TRANSFORM (DFT)

The recent advances in integrated circuit techniques are such that it is becoming more and more economical to adopt methods of signal processing in which the quantities being handled are discontinuous (sampled). A processor of sampled signals operates on batches containing a finite number of samples; accordingly a Fourier transform calculator using these techniques would be required to evaluate the discrete Fourier transform (DFT) rather than the continuous Fourier transform.

This section contains a description of the DFT and some of its properties, together with a consistent notation. The approximations involved in evaluating the Fourier transform by means of the DFT are discussed in Section 6.13. The DFT is normally evaluated by means of an efficient procedure known as the fast Fourier transform (FFT) method; this is described in Section 6.14. Finally Section 6.15 contains the proofs of various formulae quoted in Sections 6.12-6.14.

From a purely mathematical point of view, the N point DFT can be regarded as a relationship between two infinite sequences, $\{a_i\}$ and $\{A_n\}$, which have a period of N. The reference to 'a period of N' means that

$$a_i = a_{i \pm N} = a_{i \pm 2N}, \cdots$$
$$A_n = A_{n \pm N} = A_{n \pm 2N}, \cdots$$

The relationship is stated as follows: If

$$A_n = \sum_{i=0}^{N-1} a_i \, e^{-j(2\pi n/N)i} = \mathscr{D}_N\{a_i\} \qquad (6.47)$$

then

$$a_i = \frac{1}{N} \sum_{n=0}^{N-1} A_n \, e^{j(2\pi i/N)n} = \mathscr{D}_N^{-1}\{A_n\} \qquad (6.48)$$

Equations 6.47 and 6.48 define an exact relationship which is proved in Section 6.15. A_n and a_i can take either real or complex values.

6.12.1 The relationship between the DFT and the Fourier transform

A tie-up between the DFT, as defined above, and the Fourier transform can be obtained by considering the Fourier transform of an infinite train of delta functions. The relationship is derived in Section 6.15.3 and illustrated in Fig. 6.1. It can be seen that if the time delta function strengths follow the sequence a_i, the Fourier transform is another delta function train having strengths proportional to the sequence A_n.

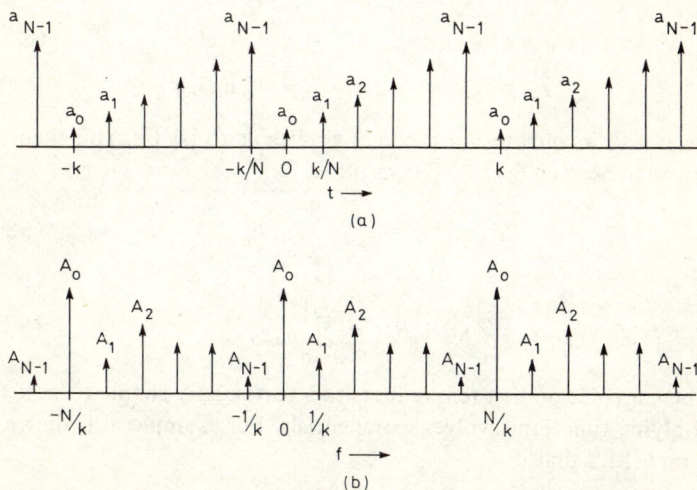

Fig. 6.1. (a) the time function $f_m(t)$*; (b) a representation of* $kF_m(j\omega)$

The Fig. 6.1 relationship can be written in the form

$$\mathscr{F}\left\{\sum_{i=0}^{N-1} a_i \delta_k\left(t - \frac{ik}{N}\right)\right\} = \frac{1}{k}\sum_{n=0}^{N-1} A_n \delta_{N/k}\left(f - \frac{n}{k}\right) \qquad (6.49)$$

where

$$\delta_k(t) = \sum_{m=-\infty}^{\infty} \delta(t - mk) \qquad (6.50)$$

If it is understood that delta functions are 'attached' to the sequence terms, and if the group repetition period (k) is unity, then the sequence $\{A_n\}$ can be regarded as the Fourier transform of the sequence $\{a_i\}$.

6.12.2 Properties of the DFT

The physical significance of the DFT can be used to obtain many DFT properties straight from known Fourier transform properties. For example, it is clear that the form of $kF_m(j\omega)$, illustrated by Fig. 6.1b, is purely symbolic. If the values a_i are real then $f_m(t)$ must be a real function of time; this means that $F_m(-j\omega)$ is equal to $F_m^*(j\omega)$ (Equation 6.7). Using this information in conjunction with Fig. 6.1b, it follows that

$$A_{-n} = A_{N-n} = A_n^*, \quad \text{for } a_i \text{ real} \qquad (6.51)$$

Other necessary and sufficient conditions, based on the results of Section 6.6, are

$$A_{-n} = A_{N-n} = -A_n^*, \quad \text{for } a_i \text{ imaginary} \qquad (6.52)$$

$$a_{-i} = a_{N-i} = a_i^*, \quad \text{for } A_n \text{ real} \qquad (6.53)$$

$$a_{-i} = a_{N-i} = -a_i^*, \quad \text{for } A_n \text{ imaginary} \qquad (6.54)$$

There is also a convolution theorem applicable to DFT's. This theorem is proved in Section 6.15.2. The result is

$$\mathscr{D}_N\{a_i b_i\} = \frac{1}{N}\sum_{m=0}^{N-1} A_m B_{n-m} \qquad (6.55)$$

$$\mathscr{D}_N^{-1}\{A_n B_n\} = \sum_{m=0}^{N-1} a_m b_{i-m} \qquad (6.56)$$

The convolution theorem reduces to a particularly simple form if the multiplying function involves exponentials. For example, it is shown in Section 6.15.2 that if

$$b_i = \cos\left(\frac{2\pi i}{N}\right) \qquad (6.57)$$

then

$$B_n = N/2, \qquad n = 1, \qquad 1 \pm N, \ldots$$
$$B_n = N/2, \qquad n = -1, \qquad -1 \pm N, \ldots \tag{6.58}$$
$$B_n = 0, \qquad \text{otherwise}$$

Hence, if

$$c_i = a_i \cos\left(\frac{2\pi i}{N}\right) \tag{6.59}$$

then

$$C_n = \tfrac{1}{2}A_{n-1} + \tfrac{1}{2}A_{n+1} \tag{6.60}$$

Thus, in the case of a time weighted function of the form discussed in Section 5.2.1,

$$c_i = a_i \left[1 + 2\beta \cos\left(\frac{2\pi i}{N}\right)\right] \tag{6.61}$$

Hence

$$C_n = A_n + \beta A_{n-1} + \beta A_{n+1} \tag{6.62}$$

showing that the effects of time weighting can be obtained by spectral processing, if desired.

6.13 EVALUATION OF THE FOURIER TRANSFORM BY MEANS OF THE DFT

It is shown below that if the sequence $\{a_i\}$ is formed from samples of $f(t)$ taken τ seconds apart then, under certain conditions, $F(j\omega)$ is represented by the DFT sequence $\{A_n\}$. The necessary conditions are

(1) $1/\tau$ must be greater than the total (i.e. +ve and −ve frequency) bandwidth (in Hz) of $F(j\omega)$.
(2) The number of samples N must be such that the truncated time function, of length $(N-1)\tau$ seconds, has essentially the same spectrum as $f(t)$.

For the usual case of a real time signal, bandlimited to B Hz, the total bandwidth is $2B$ Hz and (1) becomes $1/\tau > 2B$, i.e. the normal sampling condition.

The implications of (2), from the radar point of view, can be obtained from Sections 4.2 and 4.3 by noting that the signal processing time is $(N-1)\tau$ seconds.

If the above conditions are satisfied then $F(j\omega)$ can be evaluated for N values of ω by using Equations 6.67 and 6.68 in conjunction with Fig. 6.2.

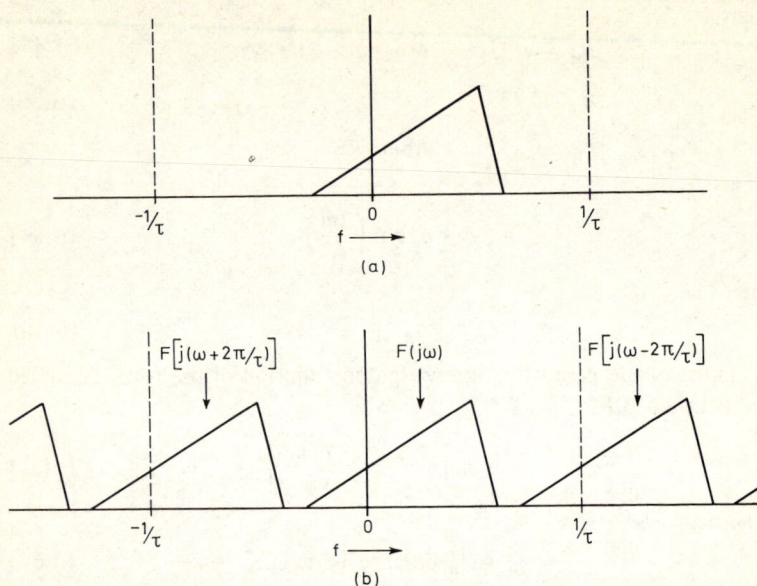

Fig. 6.2. (a) the form of F(jω) in Equation 6.64; (b) the form of τF$_s$(jω) in Equation 6.64

In a similar manner, the inverse Fourier transform of $F(j\omega)$ is represented by the sequence $\{a_i\}$ if the sequence $\{A_n\}$ is formed from samples of $F(j\omega)$ taken Δ Hz apart. The necessary conditions are

(1) $1/\Delta$ must be greater than the total duration of $f(t)$.

(2) The number of samples (N) must be such that the truncated spectrum, of width $(N - 1)\Delta$ Hz, has essentially the same inverse Fourier transform as $F(j\omega)$.

The result is summarised by Equations 6.73 and 6.74.

6.13.1 Derivation of Fourier transform results

In the treatment which follows sampled signals are represented by trains of delta functions whereas real sampled systems employ narrow pulses. It can easily be shown that sampling with a pulse of width d seconds is equivalent to impulsive sampling, provided that interest is concentrated at frequencies much lower than $1/d$ Hz and that a strength of d (rather than unity) is allocated to the sampling delta functions. The latter modification also serves to restore the correct dimensions (the dimensions of the delta function are time^{-1}).

Denoting the impulsively sampled form of $f(t)$ by $f_s(t)$, it follows that

$$f_s(t) = f(t) \sum_{i=-\infty}^{\infty} \delta(t - i\tau)$$

$$= f(t) \frac{1}{\tau} \sum_{i=-\infty}^{\infty} e^{j(2\pi i/\tau)t} \qquad (6.63)$$

The second half of Equation 6.63 follows from Equation 6.46. Hence the Fourier transform is given by

$$F_s(j\omega) = \frac{1}{\tau} \sum_{i=-\infty}^{\infty} F\left[j\left(\omega - \frac{2\pi i}{\tau}\right)\right] \qquad (6.64)$$

Equation 6.64 is illustrated by Fig. 6.2 for the general case of $f(t)$ complex. It can be seen that provided $1/\tau$ is greater than the total bandwidth (Hz) of $F(j\omega)$, no spectral overlap occurs in $F_s(j\omega)$. Hence $F(j\omega)$ can be obtained by evaluating $\tau F_s(j\omega)$.

If, for an appropriate value of time displacement (t_0), and large enough N, $f_s(t)$ can be represented by

$$f_s(t) \simeq f(t) \sum_{i=0}^{N-1} \delta(t - t_0 - i\tau) \qquad (6.65)$$

Then

$$F_s(j\omega) \simeq \int_{-\infty}^{\infty} f(t) \sum_{i=0}^{N-1} \delta(t - t_0 - i\tau) \, e^{-j\omega t} \, dt$$

$$= \sum_{i=0}^{N-1} f(t_0 + i\tau) \, e^{-j\omega t_0} \, e^{-j\omega i\tau} \qquad (6.66)$$

Hence

$$F_s\left(j \frac{2\pi n}{N\tau}\right) \simeq e^{-j(2\pi n t_0/N\tau)} \sum_{i=0}^{N-1} f(t_0 + i\tau) \, e^{-j(2\pi n/N)i}$$

Thus, if

$$a_i = f(t_0 + i\tau) \qquad (6.67)$$

then

$$\tau F_s\left(j \frac{2\pi n}{N\tau}\right) \simeq e^{-j(2\pi n t_0/N\tau)} \, \tau A_n \qquad (6.68)$$

Note that it is often possible to choose the time origin such that t_0 is zero.

6.13.2 Derivation of inverse Fourier transform results

A similar procedure can be carried out for the case of the inverse Fourier transform. By re-arranging Equations 6.34 and 6.43 it follows that if

$$F_r(j\omega) = F(j\omega) \sum_{n=-\infty}^{\infty} \delta(\omega - 2\pi n\Delta) \tag{6.69}$$

then

$$f_r(t) = \frac{1}{2\pi\Delta} \sum_{n=-\infty}^{\infty} f\left(t - \frac{n}{\Delta}\right) \tag{6.70}$$

The units of Δ are Hz.

The relationship given by Equation 6.70 is of the same form as that illustrated by Fig. 6.2, showing that $f(t)$ can be obtained by evaluating $2\pi\Delta\, f_r(t)$, provided that $1/\Delta$ is greater than the total duration of $f(t)$. The same reasoning used for the time function case shows that if, for an appropriate value of ω_0 and large enough N, $F_r(j\omega)$ can be represented by

$$F_r(j\omega) \simeq F(j\omega) \sum_{n=0}^{N-1} \delta(\omega - \omega_0 - 2\pi n\Delta) \tag{6.71}$$

then

$$f_r\left(\frac{i}{N\Delta}\right) \simeq e^{j(\omega_0 i/N\Delta)} \frac{1}{2\pi} \sum_{n=0}^{N-1} F[j(\omega_0 + 2\pi n\Delta)]\ e^{j(2\pi i/N)n} \tag{6.72}$$

Thus, if

$$A_n = F[j(\omega_0 + 2\pi n\Delta)] \tag{6.73}$$

then

$$2\pi\Delta f_r\left(\frac{i}{N\Delta}\right) \simeq e^{j(\omega_0 i/N\Delta)} N\Delta a_i \tag{6.74}$$

6.14 THE FAST FOURIER TRANSFORM (FFT) PROCESS

The fast Fourier transform (FFT) process is an efficient method of evaluating the discrete Fourier transform; it is based upon two theorems which are derived in Section 6.15.4.

Reference to the DFT definition relationship (Equation 6.47) shows that N operations are required to calculate a single value of the sequence $\{A_n\}$. Thus it would appear that N^2 operations are required to calculate the full N values of the sequence. However, it is not always necessary to carry out N^2 operations. It is shown in Section 6.15.4 (Equations 6.95 and 6.96) that, if N is an even number

$$A_{2r} = \mathscr{D}_{N/2}\{a_i + a_{i+N/2}\} \tag{6.75}$$

$$A_{2r+1} = \mathscr{D}_{N/2}\{(a_i - a_{i+N/2})e^{-j(2\pi i/N)}\} \tag{6.76}$$

Equations 6.75 and 6.76 are illustrated by Fig. 6.3, for the case of $N = 8$. The multiplier W is equal to $\exp(-j2\pi/N)$. It can be seen that the original $N^2 = 64$ operations has been reduced to $2 \times (\frac{1}{2}N)^2 + N = 40$ operations; furthermore there is no reason why the two 4-point DFT's should be evaluated by the direct method. If the FFT process is repeated each 4-point DFT will need 12 operations rather than 16. Thus the number of operations could be reduced from 64 to a total of 32. The saving in processing time becomes more dramatic as N is increased. If $N = 2^m$ the FFT process needs a total of $(m + 1)N$ operations, rather than N^2.

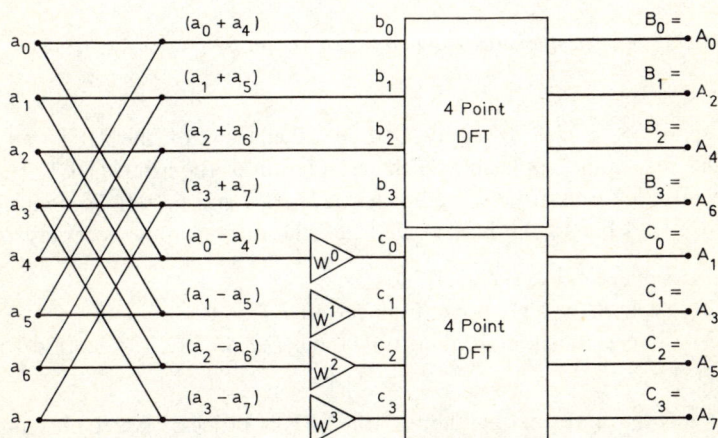

Fig. 6.3. *The reduction of an 8-point DFT to two 4-point DFT's* $W = \exp(-j2\pi/8)$

An alternative procedure can be obtained using Equations 6.99-6.101.

$$\mathscr{D}_N\{a_i\} = \mathscr{D}_{N/2}\{a_{2i}\} + e^{-j(2\pi n/N)}\mathscr{D}_{N/2}\{a_{2i+1}\} \tag{6.77}$$

Hence, if $b_i = a_{2i}$ and $c_i = a_{2i+1}$, then

$$A_m = B_m + e^{-j(2\pi m/N)}C_m, \qquad 0 \leqslant m \leqslant N/2-1 \tag{6.78}$$

$$A_{m+N/2} = B_m - e^{-j(2\pi m/N)}C_m, \qquad 0 \leqslant m \leqslant N/2-1 \tag{6.79}$$

The above relationships (for the case $N = 8$) are illustrated by Fig. 6.4. It can be seen that this method is a 'reverse' form of the first method.

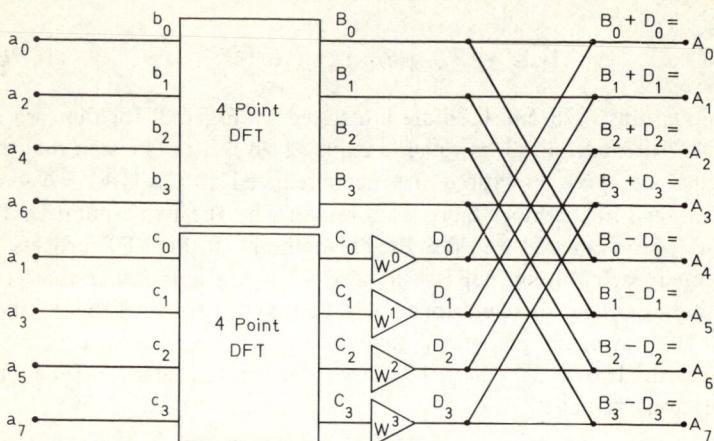

Fig. 6.4. An alternative reduction of an 8-point DFT $W = \exp(-j2\pi/8)$

Equations 6.75-6.79 apply to the calculation of the DFT; very similar results are applicable to the calculation of the inverse DFT. The relevant equations are 6.97, 6.98, 6.102, 6.103 and 6.104. To convert Figs 6.3 and 6.4 to the inverse DFT calculation it is only necessary to

(1) Replace a_i by A_i and A_i by $2\,a_i$.
(2) Replace the DFT boxes by inverse DFT boxes.
(3) Change the W multiplier to $W = \exp(j2\pi/N)$, i.e. $W = \exp(j2\pi/8)$, for $N = 8$.

For further information about the FFT process see Gold and Rader [9].

6.15 PROOFS OF THE PROPERTIES OF THE DFT AND FFT

The formal proofs of the DFT transform property and the DFT convolution theorem depend upon the following relationship

$$\sum_{m=0}^{N-1} e^{j(2\pi k/N)m} = \left.\begin{array}{l} N, \quad k = 0, \pm N, \pm 2N, \text{ etc.} \\[6pt] 0, \quad k = \text{any other integer} \end{array}\right\} \quad (6.80)$$

In the first case the argument of the summation is 1 for all m, hence the sum is clearly N. In the second case the summation is a geometric progression and it follows that

$$\sum_{m=0}^{N-1} e^{j(2\pi k/N)m} = \frac{1 - e^{j2\pi k}}{1 - e^{j(2\pi k/N)}}$$

which is equal to zero, for $k \neq 0, \pm N$, etc.

6.15.1 The basic DFT property

Using the definition of the DFT given in Section 6.12, one can write

$$\mathscr{D}_N^{-1}\{\mathscr{D}_N(a_i)\} = \frac{1}{N} \sum_{n=0}^{N-1} \sum_{r=0}^{N-1} a_r e^{-j(2\pi n/N)r} e^{j(2\pi i/N)n}$$

$$= \frac{1}{N} \sum_{r=0}^{N-1} a_r \sum_{n=0}^{N-1} e^{j[2\pi(i-r)/N]n} \qquad (6.81)$$

Equation 6.80 shows that the inner summation is zero unless $(i - r)$ is equal to $0, \pm N$, etc. Hence Equation 6.81 becomes

$$\mathscr{D}_N^{-1}\{\mathscr{D}_N(a_i)\} = a_i \qquad (6.82)$$

It also follows that $a_i = a_{i+N} = a_{i+2N}$, etc.

A similar procedure can be used to show that

$$\mathscr{D}_N\{\mathscr{D}_N^{-1}(A_n)\} = A_n \qquad (6.83)$$

Also $A_n = A_{n+N} = A_{n+2N}$, etc.

6.15.2 The DFT convolution theorem

To prove the convolution theorem

$$\frac{1}{N} \sum_{m=0}^{N-1} A_m B_{n-m} = \frac{1}{N} \sum_{m=0}^{N-1} \sum_{i=0}^{N-1} a_i e^{-j(2\pi m/N)i} \sum_{k=0}^{N-1} b_k e^{-j[2\pi(n-m)/N]k}$$

$$= \frac{1}{N} \sum_{i=0}^{N-1} \sum_{k=0}^{N-1} a_i b_k e^{-j(2\pi n/N)k} \sum_{m=0}^{N-1} e^{-j[2\pi(i-k)/N]m}$$

Hence, using Equation 6.80

$$\frac{1}{N} \sum_{m=0}^{N-1} A_m B_{n-m} = \sum_{i=0}^{N-1} a_i b_i e^{-j(2\pi n/N)i} = \mathscr{D}_N\{a_i b_i\} \qquad (6.84)$$

A similar procedure can be used to show that

$$\mathcal{D}_N^{-1}\{A_n B_n\} = \sum_{m=0}^{N-1} a_m b_{i-m} \tag{6.85}$$

For the case $b_i = \cos(2\pi i/N)$

$$B_n = \sum_{i=0}^{N-1} \tfrac{1}{2}\left[e^{j(2\pi i/N)} + e^{-j(2\pi i/N)}\right]e^{-j(2\pi n/N)i}$$

$$= \tfrac{1}{2} \sum_{i=0}^{N-1} e^{j[2\pi(1-n)/N]i} + e^{-j[2\pi(1+n)/N]i}$$

Hence, using Equation 6.80

$$B_n = \begin{cases} N/2, & n = 1, 1\pm N, \text{ etc.} \\ N/2, & n = -1, -1\pm N, \text{ etc.} \\ 0, & \text{otherwise} \end{cases}$$

Similarly, for $b_i = \sin(2\pi i/N)$

$$B_n = \begin{cases} -jN/2, & n = 1, 1\pm N, \text{ etc.} \\ jN/2, & n = -1, -1\pm N, \text{ etc.} \\ 0, & \text{otherwise} \end{cases}$$

6.15.3 The relationship between the DFT and a delta function train

The proof of the relationship between the DFT and an infinite train of delta functions will now be given. Consider a delta function train $f_m(t)$ derived from a sequence $\{a_i\}$ of N sample values. Let

$$f_m(t) = \sum_{i=0}^{N-1} a_i \delta_k\left(t - \frac{ik}{N}\right) \tag{6.86}$$

where

$$\delta_k(t) = \sum_{n=-\infty}^{\infty} \delta(t-nk) \tag{6.87}$$

It can readily be shown that

$$\mathcal{F}\{\delta_k(t)\} = \frac{2\pi}{k} \delta_{2\pi/k}(\omega) \tag{6.88}$$

Hence

$$F_m(j\omega) = \sum_{i=0}^{N-1} a_i \, e^{-j(\omega i k/N)} \frac{2\pi}{k} \delta_{2\pi/k}(\omega)$$

$$= \sum_{n=-\infty}^{\infty} \delta\left(\omega - \frac{2\pi n}{k}\right) \frac{2\pi}{k} \sum_{i=0}^{N-1} a_i \, e^{-j(2\pi n/N)i}$$

Since the i summation is periodic in n, it follows that $F_m(j\omega)$ can be written in the form

$$F_m(j\omega) = \frac{2\pi}{k} \sum_{n=0}^{N-1} A_n \delta_{2\pi N/k}\left(\omega - \frac{2\pi n}{k}\right) \tag{6.89}$$

where

$$A_n = \sum_{i=0}^{N-1} a_i \, e^{-j(2\pi n/N)i} \tag{6.90}$$

By putting $\omega = 2\pi f$, and using Equation 6.33, Equation 6.89 can be written in the form

$$F_m(j\omega) = \frac{1}{k} \sum_{n=0}^{N-1} A_n \delta_{N/k}\left(f - \frac{n}{k}\right) \tag{6.91}$$

Using the relationship

$$\mathscr{F}^{-1}\{\delta_{2\pi C}(\omega)\} = \frac{1}{2\pi C} \delta_{1/C}(t) \tag{6.92}$$

it is possible to work backwards from Equation 6.89 and say

$$f_m(t) = \mathscr{F}^{-1}\{F_m(j\omega)\} = \frac{2\pi}{k} \sum_{n=0}^{N-1} A_n \, e^{j(2\pi n/k)t} \frac{k}{2\pi N} \delta_{k/N}(t)$$

$$= \sum_{i=-\infty}^{\infty} \delta\left(t - \frac{ik}{N}\right) \frac{1}{N} \sum_{n=0}^{N-1} A_n \, e^{j(2\pi i/N)n}$$

which is equivalent to

$$f_m(t) = \sum_{i=0}^{N-1} \delta_k\left(t - \frac{ik}{N}\right) \frac{1}{N} \sum_{n=0}^{N-1} A_n \, e^{j(2\pi i/N)n} \tag{6.93}$$

If Equation 6.93 is compared with Equation 6.86 (the defining equation) it can be seen that the sample values are given by

$$a_i = \frac{1}{N} \sum_{n=0}^{N-1} A_n \, e^{j(2\pi i/N)n} \tag{6.94}$$

Thus Equations 6.90 and 6.94 form a logical definition for the DFT and its inverse.

6.15.4 Proof of the FFT theorems

The theorems leading to the FFT process will now be proved assuming that N is even.

Equation 6.90 can be written in the form

$$A_n = \sum_{i=0}^{N/2-1} a_i\, e^{-j(2\pi n/N)i} + \sum_{i=N/2}^{N-1} a_i\, e^{-j(2\pi n/N)i}$$

which, with a change of variable in the second term, becomes

$$A_n = \sum_{i=0}^{N/2-1} a_i\, e^{-j(2\pi n/N)i} + \sum_{i=0}^{N/2-1} a_{i+N/2}\, e^{-j(2\pi n/N)(i+N/2)}$$

$$= \sum_{i=0}^{N/2-1} (a_i + a_{i+N/2}\, e^{-j\pi n})\, e^{-j(2\pi n/N)i}$$

Hence

$$A_{2r} = \sum_{i=0}^{N/2-1} (a_i + a_{i+N/2})\, e^{-j[2\pi r/(N/2)]i} \tag{6.95}$$

$$A_{2r+1} = \sum_{i=0}^{N/2-1} (a_i - a_{i+N/2})\, e^{-j(2\pi i/N)}\, e^{-j[2\pi r/(N/2)]i} \tag{6.96}$$

A similar process, starting with Equation 6.94 gives

$$a_{2r} = \frac{1}{N} \sum_{n=0}^{N/2-1} (A_n + A_{n+N/2})\, e^{j[2\pi r/(N/2)]n} \tag{6.97}$$

$$a_{2r+1} = \frac{1}{N} \sum_{n=0}^{N/2-1} (A_n - A_{n+N/2})\, e^{j(2\pi n/N)}\, e^{j[2\pi r/(N/2)]n} \tag{6.98}$$

An alternative way of writing Equation 6.90 is

$$A_n = \sum_{i=0}^{N/2-1} a_{2i}\, e^{-j(2\pi n/N)2i} + \sum_{i=0}^{N/2-1} a_{2i+1}\, e^{-j(2\pi n/N)(2i+1)}$$

$$= \mathscr{D}_{N/2}\{a_{2i}\} + e^{-j(2\pi n/N)}\, \mathscr{D}_{N/2}\{a_{2i+1}\} \tag{6.99}$$

Putting $b_i = a_{2i}$ and $c_i = a_{2i+1}$, Equation 6.99 can be written in the form

$$A_m = B_m + e^{-j(2\pi m/N)} C_m, \qquad 0 \leqslant m \leqslant N/2-1 \tag{6.100}$$

$$A_{m+N/2} = B_m - e^{-j(2\pi m/N)} C_m, \qquad 0 \leqslant m \leqslant N/2-1 \tag{6.101}$$

Equation 6.101 follows since the period of B_m and C_m is $N/2$, rather than N.

As before, a similar process can be carried out on Equation 6.94 giving

$$2\,a_i = \mathscr{D}_{N/2}^{-1}\{A_{2n}\} + e^{j(2\pi i/N)}\,\mathscr{D}_{N/2}^{-1}\{A_{2n+1}\} \qquad (6.102)$$

Also, if $B_n = A_{2n}$ and $C_n = A_{2n+1}$, then

$$2\,a_m = b_m + e^{j(2\pi m/N)}\,c_m, \qquad 0 \leqslant m \leqslant N/2 - 1 \qquad (6.103)$$

$$2\,a_{m+N/2} = b_m - e^{j(2\pi m/N)}\,c_m, \qquad 0 \leqslant m \leqslant N/2 - 1 \qquad (6.104)$$

HILBERT TRANSFORMS AND COMPLEX ANALYTIC SIGNALS

When dealing with real RF signals it is mathematically convenient to work in terms of complex signals having one-sided spectra. This chapter deals with such signals and also considers the Hilbert transform which occurs in the study of their properties.

7.1 SUMMARY OF THE MAIN RESULTS OF CHAPTER 7

(1) The spectrum of a real signal consists of positive and negative frequencies which are related by the symmetry of Equation 6.7. Any spectrum not possessing this symmetry must necessarily belong to a signal which is either complex or purely imaginary.

The term 'complex analytic signal' is used in the literature of signal theory to denote a particular class of complex signals having spectra which contain no negative frequencies. In particular the complex analytic signal $f_a(t)$ corresponding to the real signal $f(t)$ is *defined* through the following property

$$\mathscr{F}\{f_a(t)\} = F_a(\mathrm{j}\omega) = F(\mathrm{j}\omega), \quad \left.\begin{array}{ll} 2F(\mathrm{j}\omega), & \omega > 0 \\ F(\mathrm{j}\omega), & \omega = 0 \\ 0, & \omega < 0 \end{array}\right\} \tag{7.1}$$

where

$$F(\mathrm{j}\omega) = \mathscr{F}\{f(t)\}$$

It is shown in Section 7.2.1 that this definition of the Fourier transform of $f_a(t)$ leads to

$$f_a(t) = f(t) + \mathrm{j}\hat{f}(t) \tag{7.2}$$

where

$$\hat{f}(t) = \frac{1}{\pi} \int_{-\infty}^{\infty} \frac{f(x)}{(t-x)} \, \mathrm{d}x \tag{7.3}$$

The integral in Equation 7.3 has to be taken in the sense of a principal value. $\hat{f}(t)$ is called the Hilbert transform of $f(t)$.

(2) Some properties of the Hilbert transform are listed and proved

in Section 7.2. It is shown in Section 7.2.2 that $\hat{f}(t)$ describes the waveform which would be obtained if $f(t)$ was passed through a perfect, broadband, $90°$ phase lag circuit. It is also shown that the real and imaginary parts of the transfer function of a physical network are related through the Hilbert transformation.

(3) The complex analytic signal concept is particularly useful in the study of RF signals of the form

$$f(t) = |a(t)| \cos[\omega_0 t + \phi(t)] \qquad (7.4)$$

where $|a(t)|$ and $\phi(t)$ are the real signals applied to the amplitude and phase modulation terminals respectively. Provided that the carrier frequency ω_0 is so high that there is negligible low frequency energy, Equation 7.2 does not have to be used to calculate $f_a(t)$; rather $f_a(t)$ can be written down by inspection as

$$f_a(t) = |a(t)| \, e^{j\phi(t)} \, e^{j\omega_0 t} = a(t) \, e^{j\omega_0 t} \qquad (7.5)$$

It should be noted that one application where Equation 7.5 may not be valid is in the study of sonar [2].

(4) Once the form of the complex analytic signal has been calculated, results obtained from operations upon such signals can be related to the corresponding real signals through various properties of the complex analytic signal.

The modulus of the complex analytic signal is a real function which corresponds to the function engineers would call the 'RF envelope'.

If a real filter processes a real signal $f(t)$ and gives a real output signal $h(t)$, the same filter would process $f_a(t)$ to give an output of $h_a(t)$.

The effect of multiplying two real signals $f(t)$, $g(t)$ is to produce an output signal having high (sum) and low (difference) frequency components. In terms of complex analytic signals the high frequency component is given by

$$\tfrac{1}{2} \operatorname{Re}\{f_a(t) g_a(t)\}$$

while the low frequency component is given by

$$\tfrac{1}{2} \operatorname{Re}\{f_a(t) g_a^*(t)\}$$

A key property of the complex analytic signal, from the point of view of the r.m.s. error criterion is that

$$\int_{-\infty}^{\infty} [f(t) - g(t)]^2 \, dt = \tfrac{1}{2} \int_{-\infty}^{\infty} |f_a(t) - g_a(t)|^2 \, dt \qquad (7.6)$$

7.2 THE HILBERT TRANSFORM

The Hilbert transform of $f(t)$ is defined as

$$\mathcal{H}\{f(t)\} = \hat{f}(t) = \frac{1}{\pi} \int_{-\infty}^{\infty} \frac{f(x)}{(t-x)}\, dx \qquad (7.7)$$

The integral has to be taken in the sense of a principal value, i.e.

$$\int_{-\infty}^{\infty} = \operatorname*{Lt}_{\lambda \to 0} \left\{ \int_{-\infty}^{t-\lambda} + \int_{t+\lambda}^{\infty} \right\}$$

This section will be used to show the significance of the Hilbert transform to the study of complex analytic signals and physical transfer functions. The following properties will also be proved

$$\mathcal{F}\{\hat{f}(t)\} = \left. \begin{array}{ll} -jF(j\omega), & \omega > 0 \\ 0, & \omega = 0 \\ jF(j\omega), & \omega < 0 \end{array} \right\} \qquad (7.8)$$

$$\hat{\hat{f}}(t) = -f(t) \qquad (7.9)$$

$$\int_{-\infty}^{\infty} f(t)\hat{g}(t)\, dt = -\int_{-\infty}^{\infty} \hat{f}(t)g(t)\, dt \qquad (7.10)$$

$$\int_{-\infty}^{\infty} \hat{f}(t)\hat{g}(t)\, dt = \int_{-\infty}^{\infty} f(t)g(t)\, dt \qquad (7.11)$$

$$\int_{-\infty}^{\infty} [f(t)-g(t)]^2\, dt = \tfrac{1}{2} \int_{-\infty}^{\infty} |f_a(t)-g_a(t)|^2\, dt \qquad (7.12)$$

7.2.1 The complex analytic signal

The relationship between the Hilbert transform and the complex analytic signal can be proved rigorously by the theory of contour integrals (Gouriet [11]). Since rigour—like beauty—is sometimes in the eye of the beholder, engineers may find the following treatment more convincing.

The Fourier transform of $f_a(t)$ is defined by Equation 7.1. This may be re-written using the unit step function (Section 6.2) as

$$\mathcal{F}\{f_a(t)\} = 2F(j\omega)u(\omega) \qquad (7.13)$$

From Equation 6.24

$$\mathcal{F}^{-1}\{u(\omega)\} = \tfrac{1}{2}\delta(t) - \frac{1}{2\pi}i(t)$$

where

$$i(t) = \begin{cases} 1/jt, & t \neq 0 \\ 0, & t = 0 \end{cases}$$

Hence $f_a(t)$ can be found from Equation 7.13 by use of the convolution theorem (Section 6.10).

$$f_a(t) = 2\mathcal{F}^{-1}\{F(j\omega)\,u(\omega)\}$$

$$= \int_{-\infty}^{\infty} f(x)\delta(t-x)\,dx - \frac{1}{\pi}\int_{-\infty}^{\infty} f(x)i(t-x)\,dx \qquad (7.14)$$

The sifting property of the delta function, Equation 6.30, shows the first integral in Equation 7.14 to be equal to $f(t)$, while the precise definition of $i(t)$ shows the second integral to be the principal value of

$$\frac{1}{\pi}\int_{-\infty}^{\infty} \frac{f(x)}{j(t-x)}\,dx = -j\hat{f}(t)$$

Hence

$$f_a(t) = f(t) + j\hat{f}(t) \qquad (7.15)$$

7.2.2 The phase shifting action of the Hilbert transform

From Equation 7.15

$$\mathcal{F}\{\hat{f}(t)\} = -j[\mathcal{F}\{f_a(t)\} - \mathcal{F}\{f(t)\}] \qquad (7.16)$$

Using Equation 7.16 with 7.13 leads to

$$\mathcal{F}\{\hat{f}(t)\} = \begin{cases} -j[2F(j\omega) - F(j\omega)], & \omega > 0 \\ -j[F(j\omega) - F(j\omega)], & \omega = 0 \\ -j[0 - F(j\omega)], & \omega < 0 \end{cases}$$

which reduces to Equation 7.8.

Since

$$j = e^{j\pi/2}, \qquad -j = e^{-j\pi/2}$$

Equation 7.8 credits the Hilbert transform with retarding the phase of all positive frequencies by $90°$ and advancing the phase of all negative frequencies by $90°$.

The above is the action of a perfect 90° phase lag network. In fact, from Appendix 4

$$\mathcal{H}\{\cos(\omega t)\} = \sin(\omega t), \qquad \omega > 0 \qquad (7.17)$$

Incidentally the above operation, yielding a real output for a real input, should not be confused with merely multiplying by j which changes a real signal into an imaginary one. Multiplying by j advances the phase of both +ve and −ve frequencies by 90° and thus destroys the essential spectral symmetry of a real signal.

7.2.3 The inverse Hilbert transform

Applying Equation 7.8 to the double Hilbert transform gives

$$\mathcal{F}\{\hat{\hat{f}}(t)\} = \begin{cases} -F(j\omega), & \omega > 0 \\ 0, & \omega = 0 \\ -F(j\omega), & \omega < 0 \end{cases} \qquad (7.18)$$

Thus Equation 7.18 leads to

$$\hat{\hat{f}}(t) = -f(t) \qquad (7.19)$$

Equation 7.19 may be regarded as an inversion relationship, i.e.

$$f(t) = -\frac{1}{\pi} \int_{-\infty}^{\infty} \frac{\hat{f}(x)}{(t-x)} \, dx \qquad (7.20)$$

The loss of d.c. component shown by Equation 7.18 is of no practical significance. If one insists on dealing with signals which possess a d.c. component and exist for all values of t from $-\infty$ to $+\infty$, the d.c. component is lost. However all practical signals are *transients*—even repetitive ones—thus the suppression of the $\omega = 0$ line will show up as a droop which will be infinitely small over any finite interval.

7.2.4 One-sided time functions

Equation 7.15 can be regarded as a way of writing the inverse Fourier transform of a one-sided spectrum. A similar result can be obtained by considering the Fourier transform of a one-sided time function. Let $f(t)$ be a real one-sided, Fourier transformable time function, specifically

$$f(t) = \begin{cases} f_1(t), & t > 0 \\ \frac{1}{2}f_1(t), & t = 0 \\ 0, & t < 0 \end{cases} \qquad (7.21)$$

A two-sided function $g(t)$ can be formed by reflecting $f(t)$ about the origin, i.e.

$$g(t) = f(t) + f(-t) \tag{7.22}$$

in other words

$$g(t) = \begin{cases} f_1(t), & t \geqslant 0 \\ f_1(-t), & t < 0 \end{cases} \tag{7.23}$$

Equations 7.22 or 7.23 allow $f(t)$ to be written

$$f(t) = g(t)u(t) \tag{7.24}$$

Since $f(t)$ is Fourier transformable, it follows from Equation 7.22 that $g(t)$ is also Fourier transformable. Also, from Equation 6.18

$$\mathscr{F}\{u(t)\} = \pi\delta(\omega) + i(\omega)$$

where

$$i(\omega) = \begin{cases} 1/j\omega, & \omega \neq 0 \\ 0, & \omega = 0 \end{cases}$$

Using the convolution theorem with Equation 7.24 gives

$$\mathscr{F}\{f(t)\} = \frac{1}{2\pi} \left[\int_{-\infty}^{\infty} G(jx)\pi\delta(\omega - x)\,dx + \int_{-\infty}^{\infty} G(jx)i(\omega - x)\,dx \right] \tag{7.25}$$

The precise definition of $i(\omega)$ shows the second integral in Equation 7.25 to be the principal value of

$$\int_{-\infty}^{\infty} \frac{G(jx)}{j(\omega - x)}\,dx$$

Hence Equation 7.25 can be written in the form

$$F(j\omega) = R(\omega) - j\hat{R}(\omega) \tag{7.26}$$

where $R(\omega) = \frac{1}{2}G(j\omega)$.

Since $f(t)$ is real it follows from Equation 7.22 and Section 6.6 (Equation 6.9) that $G(j\omega)$ is real; thus $R(\omega)$ is the real part of $F(j\omega)$ while $-\hat{R}(\omega)$ is its imaginary part. It can also be seen that $2R(\omega)$ is the Fourier transform of $f(t) + f(-t)$.

It should be stressed that $f(t)$ of Equation 7.24 must be Fourier transformable for Equation 7.26 to hold. Sections 6.3 and 6.4 show that this means that the poles of $F(p)$ must lie in the left half of the p plane. A further requirement is that the Hilbert transform of $R(\omega)$ be convergent; this means that $F(p)$ should have more poles than zeros.

7.2.5 Proof of Equations 7.10 to 7.12

To prove Equation 7.10 we write, using Equation 7.7

$$\int_{-\infty}^{\infty} f(t)\hat{g}(t)\,dt = \int_{-\infty}^{\infty} f(t)\,\frac{1}{\pi}\int_{-\infty}^{\infty}\frac{g(x)}{(t-x)}\,dx\,dt$$

The order of integration is immaterial, hence

$$\int_{-\infty}^{\infty} f(t)\hat{g}(t)\,dt = \int_{-\infty}^{\infty} g(x)\,\frac{1}{\pi}\int_{-\infty}^{\infty}\frac{f(t)}{(t-x)}\,dt\,dx \qquad (7.27)$$

The inner integral on the right-hand side of Equation 7.27 contains all terms which are functions of the variable t, and so may be evaluated separately. Comparing Equation 7.27 with Equation 7.7 gives

$$\int_{-\infty}^{\infty} f(t)\hat{g}(t)\,dt = -\int_{-\infty}^{\infty} g(x)\hat{f}(x)\,dx$$

which, since t and x are dummy variables, is equivalent to Equation 7.10.

Equation 7.11 follows directly from Equations 7.9 and 7.10, i.e.

$$\int_{-\infty}^{\infty} f(t)g(t)\,dt = -\int_{-\infty}^{\infty}\hat{\hat{f}}(t)g(t)\,dt = \int_{-\infty}^{\infty}\hat{f}(t)\hat{g}(t)\,dt$$

A corollary of Equation 7.11 is

$$\int_{-\infty}^{\infty} [\hat{f}(t)]^2\,dt = \int_{-\infty}^{\infty} [f(t)]^2\,dt \qquad (7.28)$$

Equation 7.12 can be proved using Equations 7.11, 7.15 and 7.28, i.e.

$$\int_{-\infty}^{\infty} |f_a(t)-g_a(t)|^2\,dt = \int_{-\infty}^{\infty} |[f(t)-g(t)]+j[\hat{f}(t)-\hat{g}(t)]|^2\,dt$$

The right-hand side becomes

$$\int_{-\infty}^{\infty} \{[f(t)]^2 + [g(t)]^2 - 2f(t)g(t) + [\hat{f}(t)]^2 + [\hat{g}(t)]^2 - 2\hat{f}(t)\hat{g}(t)\}\,dt$$

which, using Equations 7.11 and 7.28 reduces to

$$2\int_{-\infty}^{\infty} \{[f(t)]^2 + [g(t)]^2 - 2f(t)g(t)\}\,dt$$

proving Equation 7.12.

7.3 ENVELOPE AND PHASE FUNCTIONS

When engineers refer to the envelope of a real RF signal they mean the waveform traced out by the tips of the +ve excursions of the alternating signal. This is the waveform which is recovered by a physical envelope (or peak) detector.

A complex analytic signal has a one-sided spectrum, by definition, and can be visualised as a collection of vectors each rotating in the same direction (anti-clockwise) but with different amplitudes and frequencies. A complex analytic signal which is confined to a narrow RF band, can be visualised as a single vector rotating at a nominal carrier frequency with short term fluctuations in the speed of rotation caused by any phase modulation. The length of the carrier vector will fluctuate in accordance with any amplitude modulation.

The above visual interpretation, which is compatible with a real signal being the real part of the complex analytic signal (i.e. the projection of the rotating vector on its baseline) allows a mathematical definition of signal envelope and phase which can be applied also to the case of broadband signals where a physical definition would not be so easy. If

$$f_a(t) = f(t) + j\hat{f}(t)$$

The envelope of $f(t) = |f_a(t)| = \sqrt{[\{f(t)\}^2 + \{\hat{f}(t)\}^2]}$

The phase function of $f(t) = \text{Arg}\{f_a(t)\} = \tan^{-1}\left[\dfrac{\hat{f}(t)}{f(t)}\right]$

EXAMPLE

Let

$$f(t) = \cos(\omega_0 t)$$

Then

$$\hat{f}(t) = \sin(\omega_0 t) \text{ (from Appendix 4)}$$

Hence

$$f_a(t) = \cos(\omega_0 t) + j\sin(\omega_0 t)$$

Thus

$$\text{Envelope} = |f_a(t)| = \sqrt{\{\cos^2(\omega_0 t) + \sin^2(\omega_0 t)\}} = 1$$

The phase function $\theta(t)$ is given by

$$\text{Tan}\,[\theta(t)] = \frac{\sin(\omega_0 t)}{\cos(\omega_0 t)} = \tan(\omega_0 t)$$

Hence, the principal value of the phase function is $\omega_0 t$.

It is normally only necessary to identify a specific frequency as the carrier frequency when one wants to consider the properties of a low-pass modulating function independently of the RF signal to which it is applied. When considering a real, complex, or complex analytic low-pass modulating signal, one may define the carrier frequency as that frequency by which the low-pass spectrum has to be shifted upwards to obtain the complex analytic signal corresponding to the real modulated RF signal.

An example of the use of the above definition of carrier frequency may be given by summarising the results which will be obtained in Section 7.4.

If the spectrum of a modulating signal extends to $-f$ then the carrier frequency used with that modulating signal must be greater than f if the exponential and complex analytic signals, corresponding to the resultant real RF signal, are to be equal.

7.4 THE EXPONENTIAL APPROXIMATION TO THE COMPLEX ANALYTIC SIGNAL

Consider the complex signal

$$\alpha(t) = |a(t)|\, e^{j[\omega_0 t + \phi(t)]}$$

where $|a(t)|$, $\phi(t)$ are real arbitrary functions of time. By inspection

$$|\alpha(t)| = |a(t)|$$

$$\mathrm{Arg}[\alpha(t)] = \omega_0 t + \phi(t)$$

$$\mathrm{Re}[\alpha(t)] = |a(t)|\cos[\omega_0 t + \phi(t)] = f(t)$$

Thus, if $\alpha(t)$ were the complex analytic signal corresponding to $f(t)$, $|a(t)|$ would be the envelope of $f(t)$ and $[\omega_0 t + \phi(t)]$ would be its phase function.

Since $f(t)$ is in a form which is normally encountered in practical systems, the above satisfying results would allow one to go directly to the complex analytic signal without having to evaluate the relevant Hilbert transform.

However $\alpha(t)$, although always complex, will only be a complex analytic signal if its spectrum is zero for negative frequencies. The latter effect will only occur if the carrier frequency, ω_0, is so high that the spectrum of $f(t)$ is zero at low frequencies. If this is not so, the Hilbert transformation must be used to calculate the complex analytic signal.

To be pedantic the spectrum of $f(t)$ will never be zero at low frequencies since practical signals are not truly bandlimited; thus, strictly, the Hilbert transformation should always be used. However the errors involved in taking $\alpha(t)$ as the complex analytic signal are often small.

By way of example, consider the low frequency components of a signal consisting of a burst of carrier which has a duration of n radio frequency cycles. A study of the $(\sin x)/x$ function shows that the low frequency components will be $20 \log_{10} (n\pi)$ dB down on the radio frequency components. A 30 nS X-Band pulse would have low frequency components at least 60 dB down. In practice the attenuation would be much greater, as the $(\sin x)/x$ function applies to the spectrum of a pulse having zero rise time.

The relationship between the exponential and the complex analytic signals will now be shown. Let

$$f(t) = |a(t)| \cos[\omega_0 t + \phi(t)] \tag{7.29}$$

The corresponding exponential signal is defined as

$$\alpha(t) = |a(t)| \exp[j(\omega_0 t + \phi(t))] \tag{7.30}$$

From 7.30

$$f(t) = \text{Re}\{\alpha(t)\} = \tfrac{1}{2}[\alpha(t) + \alpha^*(t)] \tag{7.31}$$

Also, from Equation 7.15

$$f(t) = \text{Re}\{f_a(t)\} = \tfrac{1}{2}[f_a(t) + f_a^*(t)] \tag{7.32}$$

From Section 6.7

$$\mathscr{F}\{g^*(t)\} = G^*(-j\omega)$$

Thus, Equations 7.31 and 7.32 give

$$F(j\omega) = \tfrac{1}{2}\{\mathscr{A}(j\omega) + \mathscr{A}^*(-j\omega)\} \tag{7.33}$$

$$F(j\omega) = \tfrac{1}{2}\{F_a(j\omega) + [F_a^*(-j\omega)]\} \tag{7.34}$$

where

$$\mathscr{A}(j\omega) = \mathscr{F}\{\alpha(t)\}, \qquad F_a(j\omega) = \mathscr{F}\{f_a(t)\}$$

That Equations 7.33 and 7.34 do not imply $\mathscr{A}(j\omega) = F_a(j\omega)$), and hence $\alpha(t) = f_a(t)$ can be seen by reference to Figs 7.1 and 7.2. Re-arranging Equation 7.33 gives

$$\mathscr{A}(j\omega) = 2F(j\omega) - \mathscr{A}^*(-j\omega) \tag{7.35}$$

For the special case of ω_0 high enough for

$$\mathscr{A}(j\omega) = 0, \qquad \omega \leqslant 0$$

it follows that

$$\mathscr{A}(-j\omega) = 0 = \mathscr{A}^*(-j\omega), \qquad \omega \geqslant 0$$

and, from Equation 7.33

$$F(j\omega) = 0, \qquad \omega = 0$$

Hence, Equation 7.35 can be written in the form

$$\mathscr{A}(j\omega) = \begin{cases} 2F(j\omega), & \omega > 0 \\ F(j\omega), & \omega = 0 \\ 0, & \omega < 0 \end{cases}$$

which is the definition of $F_a(j\omega)$ given by Equation 7.1.

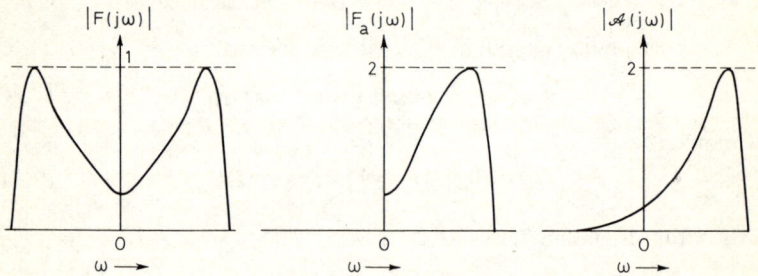

Fig. 7.1. *Spectra of* f(t) *and its related complex analytic and exponential signals*

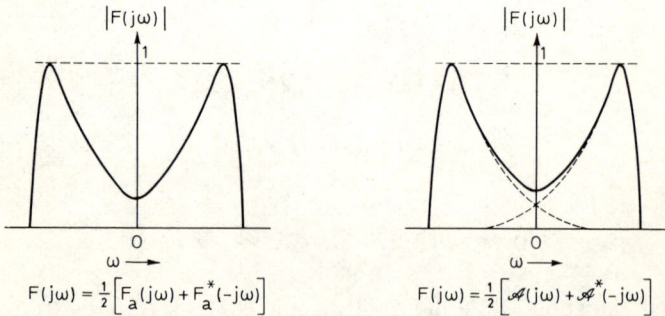

$$F(j\omega) = \tfrac{1}{2}\left[F_a(j\omega) + F_a^*(-j\omega)\right] \qquad F(j\omega) = \tfrac{1}{2}\left[\mathscr{A}(j\omega) + \mathscr{A}^*(-j\omega)\right]$$

Fig. 7.2. *The formation of the spectrum of* f(t)

Thus for the special case of a sufficiently high carrier frequency $\alpha(t) = f_a(t)$, i.e. the exponential and complex analytic signals corresponding to $f(t)$ are equal.

As a matter of interest it should be noted that if ω_0 is fixed and the above requirements are transferred to a restriction on the spectrum of the complex modulating signal, the restriction need only be applied to the −ve modulating frequencies.

7.5 THE RESULTS OF MULTIPLYING REAL OR COMPLEX ANALYTIC SIGNALS

Consider the effect of multiplying two real RF bandlimited signals $f(t)$ and $g(t)$ having spectra as indicated by Fig. 7.3. Note that

$$\omega_1 = \text{bandwidth of } f(t)$$

$$\omega_2 = \text{bandwidth of } g(t)$$

$$\omega_3 = \text{lowest frequency in spectrum of } f(t)$$

$$\omega_4 = \text{lowest frequency in spectrum of } g(t)$$

With the above notation ω_1 to ω_4 are all +ve.

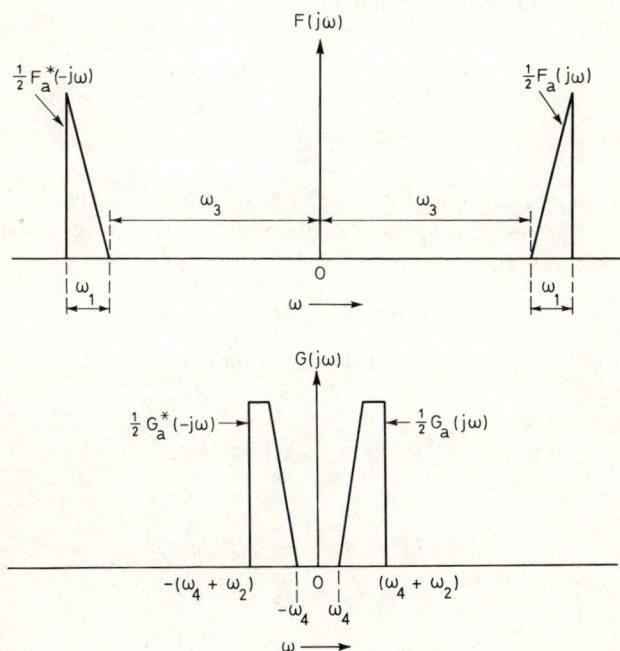

Fig. 7.3. *Spectra of real signals at the mixer input*

Experience with physical multipliers (i.e. mixers) shows that the result of such a process is the sum of a high frequency signal $h(t)$ and a low frequency signal $l(t)$. Both signals have a bandwidth of $(\omega_1 + \omega_2)$ and are separated in frequency by $(2\omega_4 - \omega_1)$.

The above statement will be proved in this section together with the facts that, (1)

$$f_a(t) g_a(t) = 2h_a(t) \tag{7.36a}$$

and hence

$$\text{Re}\{f_a(t)g_a(t)\} = 2h(t) \tag{7.36b}$$

and (2)

$$\text{Re}\{f_a(t)g_a^*(t)\} = 2l(t), \quad \text{always} \tag{7.37a}$$

$$f_a(t)g_a^*(t) = 2l_a(t), \quad \text{if } \omega_3 > (\omega_2 + \omega_4) \tag{7.37b}$$

Note that the validity condition for Equation 7.37(b) is that the lowest frequency in $f(t)$ must be greater than the highest frequency in $g(t)$.

Since the spectral separation between $l(t)$ and $h(t)$ is equal to $(2\omega_4 - \omega_1)$, the physical significance of the two signals is lost if $\omega_1 > 2\omega_4$. However, the mathematical identity of $l(t)$ and $h(t)$ is still preserved by Equations 7.36 and 7.37.

EXAMPLE

A simple demonstration of the truth of Equations 7.36 and 7.37 can be obtained by considering single sinusoids. Let $f(t) = \cos(\omega_3 t)$ and $g(t) = \cos(\omega_4 t)$. Then

$$f_a(t) = \cos(\omega_3 t) + j \sin(\omega_3 t)$$

$$g_a(t) = \cos(\omega_4 t) + j \sin(\omega_4 t)$$

Now

$$f(t)g(t) = \tfrac{1}{2}[\cos(\omega_3 + \omega_4)t + \cos(\omega_3 - \omega_4)t]$$

giving

$$h(t) = \tfrac{1}{2} \cos(\omega_3 + \omega_4)t$$

$$l(t) = \tfrac{1}{2} \cos(\omega_3 - \omega_4)t$$

Thus

$$h_a(t) = \tfrac{1}{2}[\cos(\omega_3 + \omega_4) + j \sin(\omega_3 + \omega_4)t]$$

This can be seen to be equal to $\tfrac{1}{2}f_a(t)g_a(t)$. Also

$$f_a(t)g_a^*(t) = \cos(\omega_3 - \omega_4)t + j \sin(\omega_3 - \omega_4)t$$

giving

$$\text{Re}\{f_a(t)g_a^*(t)\} = 2l(t)$$

Note that $f_a(t)g_a^*(t)$ is only equal to $2l_a(t)$ if $\omega_3 > \omega_4$, this is because $\mathcal{H}\{\cos(\omega t)\} \neq \sin(\omega t)$ if $\omega < 0$.

7.5.1 Proof of Equations 7.36 and 7.37

The general proof of the above results follows from the convolution theorem

$$\mathscr{F}\{f(t)g(t)\} = \frac{1}{2\pi} \int_{-\infty}^{\infty} F(jx)G[j(\omega-x)]\, dx$$

Expressing $F(j\omega)$, $G(j\omega)$ in terms of complex analytic spectra, this becomes

$$\mathscr{F}\{f(t)g(t)\} = \frac{1}{8\pi} \int_{-\infty}^{\infty} \{F_a(jx)+F_a^*(-jx)\}\{G_a[j(\omega-x)]+G_a^*[j(x-\omega)]\}dx$$

Performing the multiplication yields

$$\mathscr{F}\{f(t)g(t)\} = \frac{1}{8\pi} \int_{-\infty}^{\infty} \{F_a(jx)G_a[j(\omega-x)]+F_a^*(-jx)G_a^*[j(x-\omega)]\}\, dx$$

$$+ \frac{1}{8\pi} \int_{-\infty}^{\infty} \{F_a(jx)G_a^*[j(x-\omega)]+F_a^*(-jx)G_a[j(\omega-x)]\}\, dx$$

$$(7.38)$$

For convenience of reference the first integral of Equation 7.38 will be defined as $H(j\omega)$ and the second integral as $L(j\omega)$. Thus Equation 7.38 becomes

$$\mathscr{F}\{f(t)g(t)\} = H(j\omega) + L(j\omega) \qquad (7.39)$$

If, in Equation 7.38, $-\omega$ is substituted for ω and the dummy variable is changed to $-x$ it is clear that

$$H(-j\omega) = H^*(j\omega)$$

$$L(-j\omega) = L^*(j\omega)$$

Thus from Section 6.6, both $H(j\omega)$ and $L(j\omega)$ are Fourier transforms of real signals.

The physical significance of $H(j\omega)$ and $L(j\omega)$ will now be shown. Using the convolution theorem

$$\mathscr{F}\{f_a(t)g_a(t)\} = \frac{1}{2\pi} \int_{-\infty}^{\infty} F_a(jx)G_a[j(\omega-x)]\, dx \qquad (7.40)$$

$$\mathscr{F}\{f_a(t)g_a^*(t)\} = \frac{1}{2\pi} \int_{-\infty}^{\infty} F_a(jx)G_a^*[j(x-\omega)]\, dx \qquad (7.41)$$

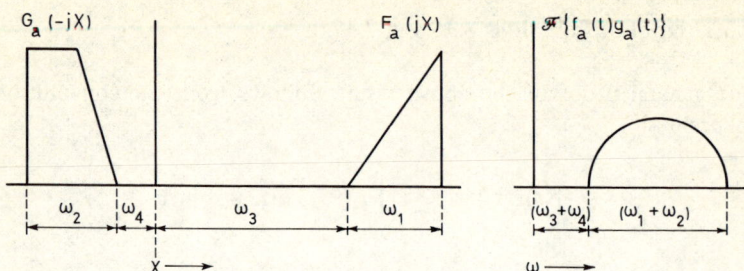

Fig. 7.4. Spectra involved in the multiplication $f_a(t)g_a(t)$. Note that $G_a[j(\omega-x)]$ is $G_a(-jx)$ moved to the right, for $\omega > 0$

It can be seen from Fig. 7.4 that Equation 7.40 defines the spectrum of a complex analytic signal (since $\omega_3 + \omega_4 > 0$). Also

$$\text{Re}\{f_a(t)g_a(t)\} = \tfrac{1}{2}[f_a(t)g_a(t) + f_a^*(t)g_a^*(t)]$$

Hence

$$\mathscr{F}\{\text{Re}[f_a(t)g_a(t)]\} = \frac{1}{4\pi}\left\{ \int\limits_{-\infty}^{\infty} F_a(jx)G_a[j(\omega-x)]\,dx \right.$$

$$\left. + \int\limits_{-\infty}^{\infty} F_a^*(-jx)G_a^*[j(x-\omega)]\,dx \right\}$$

Comparing the above with the first integral in Equation 7.38 shows

$$\mathscr{F}\{\text{Re}[f_a(t)g_a(t)]\} = 2H(j\omega)$$

The above results, together with the complex analytic nature of $f_a(t)g_a(t)$, establishes the truth of Equations 7.36(a) and (b). The proof of Equation 7.37(a) follows the same lines.

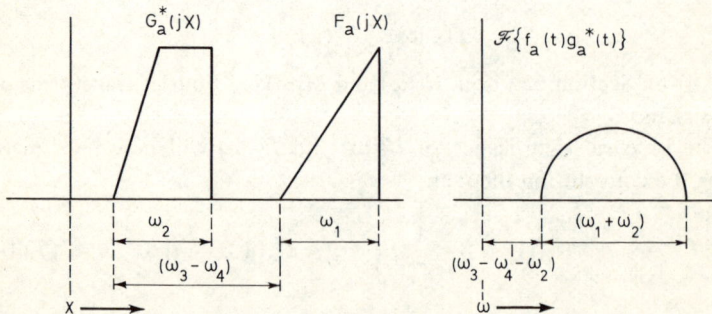

Fig. 7.5. Spectra involved in the multiplication $f_a(t)g_a^*(t)$. Note that $G_a^*[j(x-\omega)]$ is $G_a^*(jx)$ moved to the right for $\omega > 0$

Figure 7.5 shows that while $\mathscr{F}\{f_a(t)g_a^*(t)\}$ defines an asymmetric spectrum which is lower in frequency than $\mathscr{F}\{f_a(t)g_a(t)\}$, it is only one-sided if $(\omega_3 - \omega_4 - \omega_2) > 0$. This establishes Equation 7.37(b).

7.6 FILTERING EITHER REAL OR COMPLEX ANALYTIC SIGNALS

When working in terms of complex analytic signals it is essential to know how the results obtained relate to the corresponding operations using real signals. For the purposes of illustrating the effect of filtering it is convenient to define two hypothetical signal processors, shown in Fig. 7.6. The ASF (analytic signal former) converts a real signal into its complex analytic counterpart, while the RSF (real signal former) performs the inverse operation.

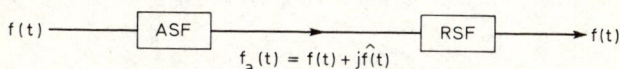

$$f(t) \longrightarrow \boxed{\text{ASF}} \longrightarrow \boxed{\text{RSF}} \longrightarrow f(t)$$
$$f_a(t) = f(t) + j\hat{f}(t)$$

Fig. 7.6. Hypothetical signal processors

It is also convenient to define the complex analytic transfer function $G_a(j\omega)$ corresponding to a physical transfer function $G(j\omega)$. By definition

$$G_a(j\omega) = \begin{cases} 2G(j\omega), & \omega > 0 \\ G(j\omega), & \omega = 0 \\ 0, & \omega < 0 \end{cases} \tag{7.42}$$

It therefore follows from Equation 7.15 that the relevant impulse responses are related by

$$g_a(t) = g(t) + j\hat{g}(t) \tag{7.43}$$

The above definitions have been used in Fig. 7.7 to present some equivalent systems. In the case of Figs 7.7(b) and (c) it can be seen that the spectrum applied to the RSF input is

$$H_a(j\omega) = \begin{cases} 2F(j\omega)G(j\omega), & \omega > 0 \\ F(j\omega)G(j\omega), & \omega = 0 \\ 0, & \omega < 0 \end{cases}$$

Hence equivalence is proved.

The RSF input spectrum in the case of Fig. 7.7(d) is

$$\begin{array}{ll} 2F(j\omega)G(j\omega), & \omega > 0 \\ \tfrac{1}{2}F(j\omega)G(j\omega), & \omega = 0 \\ 0, & \omega < 0 \end{array}$$

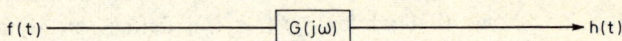

(a) Real signal into real filter

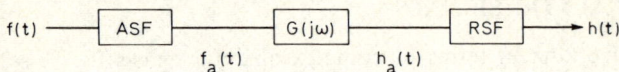

(b) Analytic signal into real filter

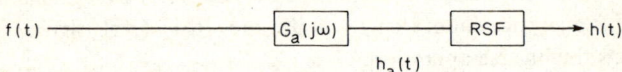

(c) Real signal into analytic filter

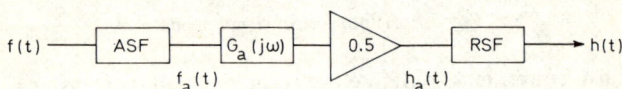

(d) Analytic signal into analytic filter

Fig. 7.7. Some equivalent systems (a) real signal into real filter (b) analytic signal into real filter (c) real signal into analytic filter (d) analytic signal into analytic filter

Although this differs from $H_a(j\omega)$ in the infinitely narrow band surrounding $\omega = 0$, the discussion in Section 7.2.3 (following Equation 7.20), coupled with the fact that the complex analytic signal method is normally used only for RF signals, means that Fig. 7.7(d) is equivalent to Figs 7.7(a), (b) and (c) for all practical purposes.

EXAMPLE

As a simple example of the use of Fig. 7.7, consider a sinusoid applied to an arbitrary filter. Let

$$f(t) = \cos(\omega_0 t), \qquad G(j\omega) = |G(j\omega)| \, e^{j\phi(j\omega)}$$

$$\hat{f}(t) = \sin(\omega_0 t), \qquad \text{from Appendix 4}$$

$$\therefore \quad f_a(t) = \cos(\omega_0 t) + j\sin(\omega_0 t) = e^{j\omega_0 t}$$

From Fig. 7.7(b) and the convolution theorem

$$h_a(t) = \int_{-\infty}^{\infty} g(x)\, e^{j\omega_0(t-x)}\, dx$$

$$= e^{j\omega_0 t} \int_{-\infty}^{\infty} g(x)\, e^{-j\omega_0 x}\, dx$$

Hence

$$h_a(t) = e^{j\omega_0 t} G(j\omega_0) = e^{j\omega_0 t} |G(j\omega_0)|\, e^{j\phi(j\omega_0)}$$

$$\therefore \quad h(t) = \mathrm{Re}\{h_a(t)\}$$

$$= |G(j\omega_0)|\, \cos[\omega_0 t + \phi(j\omega_0)]$$

as expected.

7.7 THE COMPRESSION OF A DOPPLER SHIFTED PULSE

It is a common practice in sophisticated radar systems to transmit a pulse having a long envelope—allowing high mean power—with some form of phase modulation. The uncertainty function of such a waveform often shows a potential for high precision, if the radiated bandwidth is wide enough.

One method of processing such a pulse, so as to exploit the inherent precision, is to filter it in such a fashion that the wide envelope is converted into a short envelope. The physical implications of this method are discussed in Section 5.2.2; it is the purpose of this section to develop the necessary mathematics.

Consider a real, Doppler shifted, RF signal

$$f(t) = |a(t)| \cos[(\omega_0 + \omega_d)t + \phi(t)] \tag{7.44}$$

which is filtered by a real filter, centred on the carrier frequency ω_0. The filter transfer function will be defined as

$$G(j\omega) = \tfrac{1}{2}\{B[j(\omega - \omega_0)]\, e^{j\theta[j(\omega - \omega_0)]}$$
$$+ [B[j(-\omega - \omega_0)]\, e^{j\theta[j(-\omega - \omega_0)]}]*\}$$

It is assumed that ω_0 is high enough for the complex analytic version of $G(j\omega)$ to be

$$G_a(j\omega) = B[j(\omega - \omega_0)]\, e^{j\theta[j(\omega - \omega_0)]} \tag{7.45}$$

Equation 7.45 implies that the response of the filter about ω_0 is the same as is a baseband filter

$$B(j\omega)\, e^{j\theta(j\omega)} \tag{7.46}$$

about $\omega = 0$. In the normal practical case the exponential term in expression 7.46 represents some ideal dispersive characteristic which is required for pulse compression, while $B(j\omega)$ represents the deficiencies of a practical dispersive network together with an additional bandwidth weighting term which is required for noise and sidelobe reduction.

Equation 7.44 is written in the form mainly used in this book. It is convenient, for the purposes of this section, to call the real envelope function $e(t)$ rather than $|a(t)|$. Thus $f(t)$ becomes

$$f(t) = e(t) \cos[(\omega_0 + \omega_d)t + \phi(t)] \qquad (7.47)$$

Also ω_0 is assumed high enough for

$$f_a(t) = e(t)\, e^{j\phi(t)}\, e^{j\omega_0 t}\, e^{j\omega_d t} \qquad (7.48)$$

A final piece of special notation is to call the phase modulation term $p(t)$ instead of $\exp[j\phi(t)]$, giving

$$f_a(t) = e(t)p(t)\, e^{j\omega_0 t}\, e^{j\omega_d t} \qquad (7.49)$$

With the above notation and using Section 7.6 (Fig. 7.7(d)), the output complex analytic spectrum can be written in the form

$$\mathscr{F}\{h_a(t)\} = \tfrac{1}{2}\mathscr{F}\{e(t)p(t)\, e^{j\omega_0 t}\, e^{j\omega_d t}\}B[j(\omega - \omega_0)]\, e^{j\theta\,[j(\omega - \omega_0)]} \qquad (7.50)$$

The envelope of the real output signal is then given by $|h_a(t)|$.

The rest of this section will be used to show

(1) In the general case

$$h_a(t) = \frac{1}{8\pi^2} \int_{-\infty}^{\infty} E(jx) \int_{-\infty}^{\infty} B[j(\omega - \omega_0)]\, e^{j\theta\,[j(\omega - \omega_0)]}$$

$$\times\, P[j(\omega - \omega_0 - \omega_d - x)]\, e^{j\omega t}\, d\omega\, dx \qquad (7.51)$$

(2) In the special case of linear FM, i.e.

$$p(t) = e^{jbt^2}$$

it is desirable to choose $\theta(j\omega)$ such that

$$\theta(j\omega) = \frac{\omega^2}{4b}$$

whereupon Equation 7.51 reduces to

$$h_a(t) = \frac{1}{4\pi}\, e^{j\pi/4}\, e^{j\omega_0 t}\, \sqrt{\left(\frac{\pi}{b}\right)} \int_{-\infty}^{\infty} E(jx)b\left[\frac{x}{2b} + \frac{\omega_d}{2b} + t\right]$$

$$\times\, e^{-j[(\omega_d + x)^2/4b]}\, dx \qquad (7.52)$$

(3) For the common case of wideband linear FM, Equation 7.52 becomes

$$h_a(t) \simeq e^{j\pi/4}\, e^{j\omega_0 t}\, \sqrt{\left(\frac{\pi}{4b}\right)}\, m\left[t + \frac{\omega_d}{2b}\right] \qquad (7.53)$$

where

$$m(t) = \mathscr{F}^{-1}\left\{ B(j\omega)e\left[\frac{\omega}{2b}\right]\right\}$$

The physical implications of Equations 7.52 and 7.53 are discussed in Sections 5.2.2 and 5.2.4.

7.7.1 Proof of Equation 7.51

To show that Equation 7.50 leads to Equation 7.51

$$\mathscr{F}\{e(t)p(t)\} = \frac{1}{2\pi}\int_{-\infty}^{\infty} E(jx)\,P[j(\omega - x)]\,\mathrm{d}x$$

Hence Equation 7.50 becomes

$$\mathscr{F}\{h_a(t)\} = \frac{1}{4\pi}\, B[j(\omega - \omega_0)]\, e^{j\theta\,[j(\omega - \omega_0)]}$$

$$\times \int_{-\infty}^{\infty} E(jx)P[j(\omega - \omega_0 - \omega_d - x)]\,\mathrm{d}x \qquad (7.54)$$

$h_a(t)$ can be found from Equation 7.54 by using the Fourier transform inversion integral, i.e.

$$h_a(t) = \frac{1}{8\pi^2}\int_{-\infty}^{\infty} B[j(\omega - \omega_0)]\, e^{j\theta\,[j(\omega - \omega_0)]}$$

$$\times \int_{-\infty}^{\infty} E(jx)P[j(\omega - \omega_0 - \omega_d - x)]\,\mathrm{d}x\, e^{j\omega t}\,\mathrm{d}\omega \qquad (7.55)$$

Re-arranging Equation 7.55 into a form where all the ω variable is contained in one integral gives Equation 7.51.

7.7.2 Linear FM

For the special case of linear FM

$$P(j\omega) = \mathscr{F}\{e^{jbt^2}\} \qquad (7.56)$$

Campbell and Foster [12] (pair 760) give

$$\mathscr{F}\{e^{\mp j\pi t^2}\, e^{\pm j\pi/8}\} = e^{\pm j(\omega^2/4\pi)}\, e^{\mp j\pi/8} \qquad (7.57)$$

Hence using Equation 6.6

$$\mathscr{F}\{e^{\mp j\pi a^2 t^2}\} = \frac{1}{|a|}\, e^{\pm j(\omega^2/4\pi a^2)}\, e^{\mp j\pi/4} \qquad (7.58)$$

giving

$$P(j\omega) = \sqrt{\left(\frac{\pi}{b}\right)}\, e^{j\pi/4}\, e^{-j(\omega^2/4b)} \qquad (7.59)$$

Substituting Equation 7.59 in Equation 7.51 the inner integral becomes

$$e^{j\pi/4}\sqrt{\left(\frac{\pi}{b}\right)} \int_{-\infty}^{\infty} B[j(\omega-\omega_0)]\, e^{j\theta[j(\omega-\omega_0)]}$$

$$\times\, e^{-j[(\omega-\omega_0-\omega_d-x)^2/4b]}\, e^{j\omega t}\, d\omega \qquad (7.60)$$

The above integral may be evaluated separately as it contains all the ω variable.

The method to use in the particular case of linear FM—and possibly also in the general case—is to choose $\theta[j(\omega-\omega_0)]$ such that the non-linear (in ω) terms in the other exponential are cancelled. This leaves an integral which is of the form of a Fourier transform of a wide band function (i.e. $B(j\omega)$) and which will lead to a short time function. Accordingly, we choose

$$\theta[j(\omega-\omega_0)] = \frac{(\omega-\omega_0)^2}{4b} \qquad (7.61)$$

Substituting Equation 7.61 in 7.60 gives

$$e^{j\pi/4}\sqrt{\left(\frac{\pi}{b}\right)} \int_{-\infty}^{\infty} B[j(\omega-\omega_0)]\, e^{j[(\omega-\omega_0)(\omega_d+x)/2b]}$$

$$\times\, e^{-j[(\omega_d+x)^2/4b]}\, e^{j\omega t}\, d\omega \qquad (7.62)$$

With a change of dummy variable to $(\omega+\omega_0)$, Equation 7.62 becomes

$$e^{j\pi/4}\, e^{-j[(\omega_d+x)^2/4b]}\sqrt{\left(\frac{\pi}{b}\right)} \int_{-\infty}^{\infty} B(j\omega)\, e^{j[\omega(\omega_d+x)/2b]}\, e^{j(\omega+\omega_0)t}\, d\omega$$

$$(7.63)$$

From the definition of the Fourier transform

$$b(t) = \frac{1}{2\pi} \int_{-\infty}^{\infty} B(j\omega)\, e^{j\omega t}\, d\omega$$

Hence Equation 7.63 becomes

$$2\pi \, e^{j\pi/4} \, e^{j\omega_0 t} \, e^{-j[(\omega_d + x)^2/4b]} \sqrt{\left(\frac{\pi}{b}\right)} \, b \left[t + \frac{\omega_d}{2b} + \frac{x}{2b} \right] \quad (7.64)$$

Substituting Equation 7.64 for the inner integral in Equation 7.51 gives

$$h_a(t) = \frac{1}{4\pi} \, e^{j\pi/4} \, e^{j\omega_0 t} \sqrt{\left(\frac{\pi}{b}\right)} \int\limits_{-\infty}^{\infty} E(jx) b \left[t + \frac{\omega_d}{2b} + \frac{x}{2b} \right]$$

$$\times \, e^{-j[(\omega_d + x)^2/4b]} \, dx$$

which is Equation 7.52.

7.7.3 The case of high dispersion factor

Equation 7.52 can be evaluated numerically for specific functions. An example is given in Section 5.2.2. It is shown in Section 5.2.4 that physical considerations are often such that the exponential term in the integral of Equation 7.52 can be considered equal to unity over the range of the other two functions. This leads to a substantial simplification; Equation 7.52 becomes

$$h_a(t) \simeq \frac{1}{4\pi} \, e^{j\pi/4} \, e^{j\omega_0 t} \sqrt{\left(\frac{\pi}{b}\right)} \int\limits_{-\infty}^{\infty} E(jx) b \left[\frac{x}{2b} + \frac{\omega_d}{2b} + t \right] dx \quad (7.65)$$

If the Fourier transform and its inverse are used to express $b(t)$ in terms of $B(j\omega)$, and $E(j\omega)$ in terms of $e(t)$, the integral in Equation 7.65 becomes

$$\iint\limits_{-\infty}^{\infty} e(y) \, e^{-jxy} \, dy \, \frac{1}{2\pi} \int\limits_{-\infty}^{\infty} B(j\omega) \, e^{j[(x/2b) + (\omega_d/2b) + t]\omega} \, d\omega \, dx \quad (7.66)$$

Re-arranging expression 7.66 to put all the x variable under one integral gives

$$\int\limits_{-\infty}^{\infty} e(y) \, \frac{1}{2\pi} \int\limits_{-\infty}^{\infty} e^{j[(\omega/2b) - y]x} \, dx \int\limits_{-\infty}^{\infty} B(j\omega) \, e^{j[(\omega_d/2b) + t]\omega} \, d\omega \, dy \quad (7.67)$$

The integral in x will be recognised as $\delta[(\omega/2b) - y]$, Equation 6.32. Hence 7.67 can be written

$$\int\limits_{-\infty}^{\infty} e(y) \delta \left[\frac{\omega}{2b} - y \right] \int\limits_{-\infty}^{\infty} B(j\omega) \, e^{j[(\omega_d/2b) + t]\omega} \, d\omega \, dy \quad (7.68)$$

Carrying out the y integration in 7.68 by the use of Equation 6.30 means that 7.68 is equal to

$$\int_{-\infty}^{\infty} B(j\omega)e\left[\frac{\omega}{2b}\right] e^{j[(\omega_d/2b)+t]\omega} \, d\omega \qquad (7.69)$$

If now $M(j\omega)$ is defined as

$$M(j\omega) = B(j\omega)e\left[\frac{\omega}{2b}\right]$$

The Fourier transform inversion integral gives

$$m(t) = \frac{1}{2\pi} \int_{-\infty}^{\infty} B(j\omega)e\left[\frac{\omega}{2b}\right] e^{j\omega t} \, d\omega \qquad (7.70)$$

Substituting expression 7.69 for the integral in Equation 7.65 and using Equation 7.70 gives

$$h_a(t) \simeq e^{j\pi/4} \, e^{j\omega_o t} \sqrt{\left(\frac{\pi}{4b}\right)} \, m\left[t + \frac{\omega_d}{2b}\right]$$

which is Equation 7.53.

APPENDIX 1   COMPLEX CONJUGATE TERMINOLOGY

In this book the complex conjugate of a function or a number is denoted by a single asterisk. Thus, if

$$z = x + jy, \qquad x \text{ and } y \text{ real}$$

$$z^* = x - jy, \qquad \text{by definition}$$

Similarly, if

$$f(z) = z + jz, \qquad z \text{ complex}$$

$$f^*(z) = (z + jz)^* = z^* - jz^*, \qquad \text{by definition}$$

Choosing a real variable, t,

$$f(t) = t + jt$$

then

$$f^*(t) = t^* - jt^* = t - jt, \qquad \text{for } t \text{ real}$$

Typical examples used in the present text are

$$f(t) = e^{jbt}, \qquad f^*(t) = e^{-jbt}, \qquad \text{for } b \text{ and } t \text{ real}$$

$$F(j\omega) = \frac{a}{a + j\omega}, \qquad F^*(j\omega) = \frac{a}{a - j\omega}, \qquad \text{for } a \text{ and } \omega \text{ real}$$

The above form of notation is frequently employed by authors of engineering books (see for example James, Nichols and Phillips [16]). Authors of more theoretical books often use a different form of notation, as typified by Milne-Thomson [17].

Following the notation of Reference 17, if

$$f(t) = t + jt, \qquad t \text{ real}$$

then

$$\bar{f}(t) = t - jt, \qquad \text{by definition}$$

Since the definition has been applied to the case of a real variable, rather than to the more general case of the complex variable, the normal rules of functional notation lead to

$$f(z) = z + jz, \qquad z \text{ complex}$$

$$\bar{f}(z) = z - jz, \qquad \text{which is not the complex conjugate}$$

The complex conjugate of the given $f(z)$ is $\bar{z} - j\bar{z}$, i.e. $\bar{f}(\bar{z})$. It therefore follows that the relationship between the bar notation of [17] and the asterisk notation used here is

$$\bar{f}(\bar{z}) = f^*(z)$$

APPENDIX 2   PARSEVAL'S THEOREM

Parseval's theorem as normally quoted relates the area under the squared modulus of a time function to the area under the squared modulus of its Fourier transform.

$$\int_{-\infty}^{\infty} |a(t)|^2 \, dt = \frac{1}{2\pi} \int_{-\infty}^{\infty} |A(j\omega)|^2 \, d\omega \qquad (A2.1)$$

A more general form is

$$\int_{-\infty}^{\infty} a(t)b^*(t) \, dt = \frac{1}{2\pi} \int_{-\infty}^{\infty} A(j\omega)B^*(j\omega) \, d\omega \qquad (A2.2)$$

Equation A2.1 follows from A2.2 as the special case of $b(t) = a(t)$.

To prove Equation A2.2 the convolution theorem is used with the Section 6.7 result that $\mathscr{F}\{b^*(t)\} = B^*(-j\omega)$. Hence

$$\mathscr{F}\{a(t)b^*(t)\} = \frac{1}{2\pi} \int_{-\infty}^{\infty} A(j\omega)B^*[-j(x-\omega)] \, d\omega$$

The Fourier transform definition integral gives

$$\mathscr{F}\{a(t)b^*(t)\} = \int_{-\infty}^{\infty} a(t)b^*(t) \, e^{-jxt} \, dt$$

Equation A2.2 follows as the case $x = 0$. A similar procedure shows that

$$\int_{-\infty}^{\infty} a(t)b(t) \, dt = \frac{1}{2\pi} \int_{-\infty}^{\infty} A(j\omega)B(-j\omega) \, d\omega \qquad (A2.3)$$

APPENDIX 3   FRESNEL INTEGRALS AND THEIR RELATIONSHIP TO LINEAR FM

The Fresnel integrals $C(x)$, $S(x)$, are defined as

$$C(x) = \int_{0}^{x} \cos\left(\frac{\pi t^2}{2}\right) \, dt \qquad (A3.1)$$

$$S(x) = \int_0^x \sin\left(\frac{\pi t^2}{2}\right) dt \qquad (A3.2)$$

The above integrals cannot be expressed in closed form, but tabulated results are available [13]. $C(x)$ and $S(x)$ are sketched in Fig. A3.1, for $x > 0$.

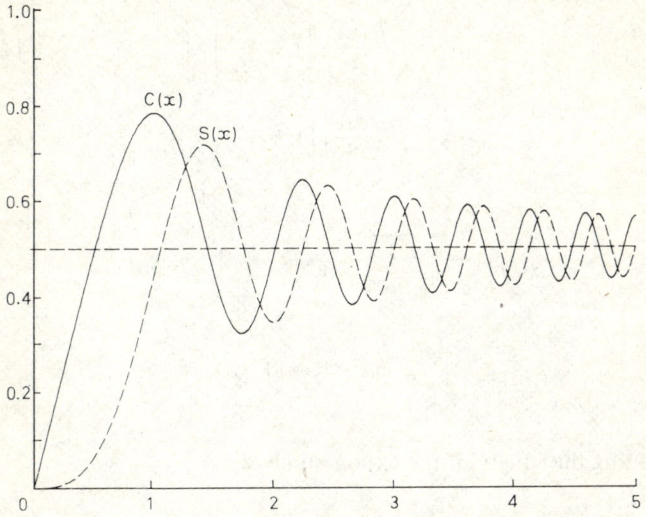

Fig. A3.1. Fresnel integrals

It can be seen from Equations A3.1 and A3.2 that

$$C(-x) = -C(x) \qquad (A3.3)$$

$$S(-x) = -S(x) \qquad (A3.4)$$

An integral which occurs in the study of linear FM can be expressed in terms of $C(x)$ and $S(x)$

$$\int_0^x e^{\pm j(\pi t^2/2)} dt = C(x) \pm jS(x) \qquad (A3.5)$$

The above results are used to show that

$$\mathcal{F}\left\{ \overset{\text{- -- 1}}{\underset{-\frac{1}{2}d \quad \frac{1}{2}d}{\rule{0pt}{0pt}}} e^{\pm jbt^2} \right\}$$

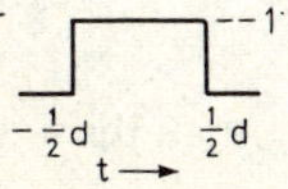

$$= e^{\mp j(\omega^2/4b)} \sqrt{\left(\frac{\pi}{2b}\right)} \{[C(u_1) + C(u_2)] \pm j[S(u_1) + S(u_2)]\} \qquad (A3.6)$$

where

$$u_1 = \sqrt{\left(\frac{bd^2}{2\pi}\right)}\left[1 \mp \frac{\omega}{bd}\right] \tag{A3.7}$$

$$u_2 = \sqrt{\left(\frac{bd^2}{2\pi}\right)}\left[1 \pm \frac{\omega}{bd}\right] \tag{A3.8}$$

Putting $b = \Delta/2d$ (see Section 2.4)

$$u_1 = \sqrt{\left(\frac{d\Delta}{4\pi}\right)}\left[1 \mp \frac{2\omega}{\Delta}\right] \tag{A3.9}$$

$$u_2 = \sqrt{\left(\frac{d\Delta}{4\pi}\right)}\left[1 \pm \frac{2\omega}{\Delta}\right] \tag{A3.10}$$

For

$$f(t) = \underset{-\frac{1}{2}d \qquad \overset{t \longrightarrow}{\frac{1}{2}d}}{\rule{2cm}{0.4pt}}\ e^{\pm jbt^2}\ , \qquad b > 0$$

$$F(j\omega) = \int_{-\frac{1}{2}d}^{\frac{1}{2}d} e^{\pm bj[t^2 \mp (\omega/b)t]}\ dt$$

Completing the square of the exponent gives

$$F(j\omega) = e^{\mp j(\omega^2/4b)} \int_{-\frac{1}{2}d}^{\frac{1}{2}d} e^{\pm bj[t \mp (\omega/2b)]^2}\ dt$$

Changing the dummy variable such that

$$y\sqrt{\left(\frac{\pi}{2}\right)} = \left(t \mp \frac{\omega}{2b}\right)\sqrt{b} \tag{A3.11}$$

gives

$$y = t\sqrt{\left(\frac{2b}{\pi}\right)} \mp \frac{\omega}{\sqrt{(2\pi b)}}$$

Hence

$$F(j\omega) = e^{\mp j(\omega^2/4b)}\sqrt{\left(\frac{\pi}{2b}\right)} \int_{t=-\frac{1}{2}d}^{t=\frac{1}{2}d} e^{\pm j(\pi y^2/2)}\ dy \tag{A3.12}$$

It follows, from Equations A3.5 and A3.11, that the integral in Equation A3.12 can be written in the form

$$C(u_1) \pm jS(u_1) - C(-u_2) \mp jS(-u_2) \tag{A3.13}$$

where

$$u_1 = d \left/ \sqrt{\left(\frac{b}{2\pi}\right)} \mp \frac{\omega}{\sqrt{(2\pi b)}} \right. \qquad \text{(A3.14)}$$

$$u_2 = d \left/ \sqrt{\left(\frac{b}{2\pi}\right)} \pm \frac{\omega}{\sqrt{(2\pi b)}} \right. \qquad \text{(A3.15)}$$

Equations A3.3 and A3.4 allow A3.13 to be written in the form

$$[C(u_1) + C(u_2)] \pm j[S(u_1) + S(u_2)] \qquad \text{(A3.16)}$$

Substituting Equation A3.16 for the integral in Equation A3.12 gives Equation A3.6.

APPENDIX 4   A SHORT TABLE OF HILBERT TRANSFORMS

The following examples are quoted in the extensive table given by Erdelyi [18].

$f(t)$	$\hat{f}(t)$
1	0
(pulse from a to b, $b > a$)	$\frac{1}{\pi} \log_e \left\| \frac{a-t}{b-t} \right\|$
$\exp(jat) \quad a > 0$	$-j \exp(jat)$
$\dfrac{\sin(at)}{t}, \quad a > 0$	$\dfrac{1 - \cos(at)}{t}$
$\cos(at), \quad a > 0$	$\sin(at)$
$-\hat{g}(t)$	$g(t)$

APPENDIX 5   A SUMMARY OF THE MAIN NOTATION

The main notation is listed below. Any deviations are indicated in the appropriate sections of the text.

Laplace transforms

$$\mathcal{L}\{f(t)\} = F(p) = \int_{-\infty}^{\infty} f(t)\, e^{-pt}\, dt, \qquad \alpha < \text{Re}(p) < \beta$$

$$\mathcal{L}^{-1}\{F(p)\} = f(t) = \frac{1}{2\pi j} \int_{c-j\infty}^{c+j\infty} F(p)\, e^{pt}\, dp, \qquad \alpha < c < \beta$$

Fourier transforms

$$\mathcal{F}\{f(t)\} = F(j\omega) = \int_{-\infty}^{\infty} f(t)\,e^{-j\omega t}dt$$

$$\mathcal{F}^{-1}\{F(j\omega)\} = f(t) = \frac{1}{2\pi} \int_{-\infty}^{\infty} F(j\omega)\,e^{j\omega t}\,d\omega$$

Hilbert transforms

$$\mathcal{H}\{f(t)\} = \hat{f}(t) = \frac{1}{\pi} \int_{-\infty}^{\infty} \frac{f(x)}{(t-x)}\,dx$$

$$\mathcal{H}^{-1}\{\hat{f}(t)\} = f(t) = -\frac{1}{\pi} \int_{-\infty}^{\infty} \frac{\hat{f}(x)}{(t-x)}\,dx$$

The above integrals are to be taken in the sense of principal values.

Fresnel integrals

$$C(x) = \int_{0}^{x} \cos\left[\frac{\pi t^2}{2}\right]dt$$

$$S(x) = \int_{0}^{x} \sin\left[\frac{\pi t^2}{2}\right]dt$$

Real functions

$$f(t), \qquad g(t), \qquad h(t)$$

Complex functions

$$a(t), \qquad b(t), \qquad c(t)$$

Complex analytic functions

$f_a(t)$ = the complex analytic function corresponding to the real $f(t)$

$$F_a(j\omega) = \mathcal{F}\{f_a(t)\}$$
$$f_a(t) = f(t) + j\hat{f}(t)$$

$$F_a(j\omega) = \begin{cases} 2F(j\omega), & \omega > 0 \\ F(j\omega), & \omega = 0 \\ 0, & \omega < 0 \end{cases}$$

General form of transmitted signal

$$f(t) = |a(t)| \cos[\omega_0 t + \phi(t)]$$

where $|a(t)|$ and $\phi(t)$ are real functions of time. The exponential signal corresponding to $f(t)$ is

$$|a(t)|\, e^{j\phi(t)}\, e^{j\omega_0 t} = a(t)\, e^{j\omega_0 t}$$

Note

$$a(t) = |a(t)|\, e^{j\phi(t)}$$

An alternative form of $a(t)$ is used in Sections 5.2 and 7.7

$$a(t) = e(t)p(t)$$

where

$$e(t) = |a(t)|$$
$$p(t) = e^{j\phi(t)}$$

Coded waveform notation

$$c_i = i\text{th (complex) code element}$$

$$\chi_b(\tau, \omega) = \chi(\tau, \omega) \text{ for a single bit pulse}$$

$$L = \text{number of bits in one code word}$$

$$\mathscr{A}(m, \omega) = \sum_{i=0}^{L-1-m} c_i c_{i+m}^* \, e^{-ji\omega d}$$

$$A(m, \omega) = \sum_{i=0}^{L-1} c_i c_{i+m}^* \, e^{-ji\omega d}$$

DFT notation

$$\mathscr{D}_N\{a_i\} = A_n = \sum_{i=0}^{N-1} a_i \, e^{-j(2\pi n/N)i}$$

$$\mathscr{D}_N^{-1}\{A_n\} = a_i = \frac{1}{N} \sum_{n=0}^{N-1} A_n \, e^{j(2\pi i/N)n}$$

Other notation

b	quadratic phase multiplier, $b = (\Delta/2d)$
c	velocity of propagation, also real part of p
d	pulse duration
E	signal energy i.e. $\displaystyle\int_{-\infty}^{\infty} [f(t)]^2 \, dt$
f	real frequency variable (Hz), $= (\omega/2\pi)$
j	$\sqrt{-1}$
k	PRF period
L	number of bits in one codeword
n	number of pulses in a repetitive train, also DFT variable
N	number of DFT samples
p	complex frequency variable, $p = (c + j\omega)$
r	range (measured with respect to target)
t	real time variable
t_v	value of Fourier transform receiver variable delay
τ	time co-ordinate (measured with respect to target)
v	velocity (measured with respect to target)
ω	real frequency variable (rad/s), $= 2\pi f$ Also Doppler co-ordinate (measured with respect to target)
ω_L	frequency of Fourier transform receiver offset oscillator
ω_0	carrier frequency (see Section 7.3)
ω_v	frequency of matched filter receiver variable oscillator
x	time co-ordinate (measured with respect to receiver), also a dummy integration variable
y	Doppler co-ordinate (measured with respect to receiver), also a dummy integration variable
z	a dummy integration variable
Δ	total frequency sweep (rad/s) of an FM pulse. $\Delta = 2bd$. In Section 6.13, Δ = distance between spectral samples (Hz)
$\delta(t)$	the delta function (see Section 6.9)
$u(t)$	the unit step function (see Sections 6.2 and 6.8)
$|\chi(\tau, \omega)|$	the uncertainty function (see Section 2)
$*$	Denotes complex conjugate (see Appendix 1)
θ	$\frac{1}{2}\omega k$
$\psi(\tau, \omega)$	the function, similar to $\chi(\tau, \omega)$, obtained from a practical receiver
$\delta_k(t)$	a train of delta functions, with period k (see Section 6.12.1)
μ	matched filter gain constant (see Section 4.4)

REFERENCES

1. WOODWARD, *Probability and Information Theory, with Application to Radar,* Pergamon Press (1955).
2. KRAMER, 'Tolerances of FM Correlation Sonars', *Proc. IEEE,* 627-36 (May 1967).
3. STUTT, 'A Note on Invariant Relations for Ambiguity and Distance Functions', *IRE Trans. Information Theory,* 164-7 (Dec. 1959).
4. STUTT, 'Some Results on Real–Part/Imaginary–Part and Magnitude–Phase Relations in Ambiguity Functions', *IEEE Trans. Information Theory,* 321-7 (Oct. 1964).
5. SIEBERT, 'Studies of Woodward's Uncertainty Function', *M.I.T. Res. Lab. of Electronics Quarterly Progress Report,* 90-4 (April 1958).
6. REKTORYS, *Survey of Applicable Mathematics,* Iliffe, 695 (1969).
7. GUILLEMIN, *The Mathematics of Circuit Analysis,* Wiley, 485-495 (1949).
8. VAN DER POL and BREMMER, *Operational Calculus,* Cambridge University Press (1964).
9. GOLD and RADER, *Digital Processing of Signals,* McGraw-Hill (1969).
10. DWIGHT, *Tables of Integrals and Other Mathematical Data,* Macmillan (1965).
11. GOURIET, 'Two Theorems Concerning Group Delay', *IEE Monograph No. 275R,* (Dec. 1957).
12. CAMPBELL and FOSTER, *Fourier Integrals for Practical Applications,* Van Nostrand (1948).
13. ABRAMOWITZ and STEGUN, *Handbook of Mathematical Functions,* Dover (1965).
14. SKOLNIK, *Introduction to Radar Systems,* McGraw-Hill, 416 (1962).
15. BLACKMAN and TUKEY, *The Measurement of Power Spectra,* Dover (1959).
16. JAMES, NICHOLS and PHILLIPS, *Theory of Servomechanisms,* McGraw-Hill, 41 (1947).
17. MILNE-THOMSON, *Theoretical Hydrodynamics,* Macmillan, 119 (1949).
18. ERDELYI, *Table of Integral Transforms,* Vol. 2, McGraw-Hill (1954).
19. LIGHTHILL, *Fourier Analysis and Generalised Functions,* Cambridge University Press (1958).

INDEX